BOSCAWEN-ÛN

BRONZE AGE
HARPEDONAPTAI
IN CORNWALL

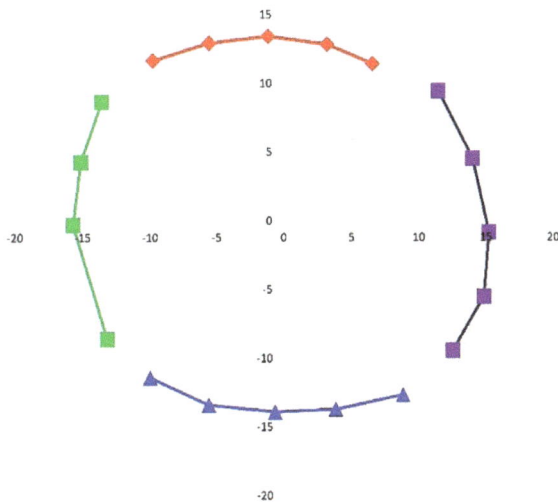

James R Warren

BLOXWICH
2023

Published in the United Kingdom in 2023 by Midland Tutorial Productions

Third Edition 30 April 2023

File Prefix Code: BOSUNed3

ISBN 978 1 915750 07 5

Midland Tutorial Productions Publishers
31 Victoria Avenue
Bloxwich
Walsall
WS3 3HS
United Kingdom

MIDLAND
TUTORIAL

BOSCAWEN-ÛN

Third Edition

James R Warren

MIDLAND TUTORIAL PRODUCTIONS
BLOXWICH

Books By James R Warren

Beyond Tourist Britain
Boscawen-Ûn (First, Second and Third Editions)
Exordium
Gamma Solution
Gleanings as I Pass
Mathematical Explorations
Meditations
Moddeshall Hydropower
Pi and Phi
Researches: Volume One
Researches: Volume Two
Researches: Volume Three
Researches: Volume Four
Unreasonable Mathematics

To The Glory of The Loving God

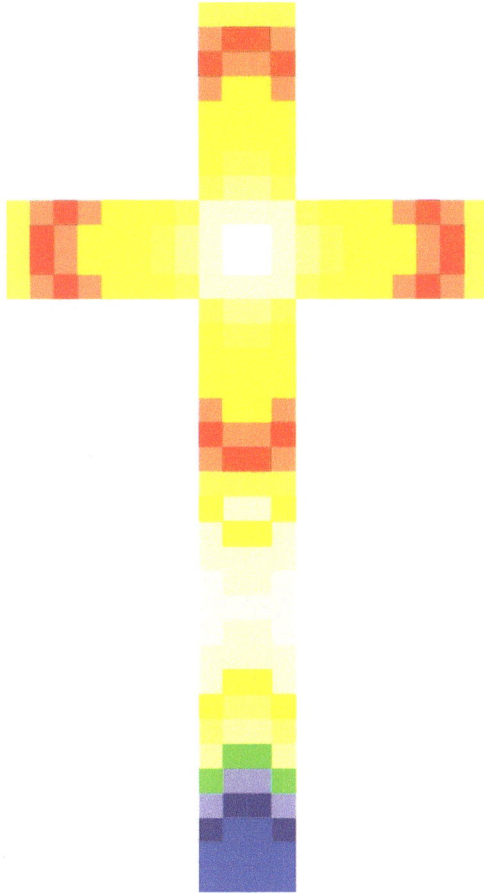

Who Made Our Minds Free

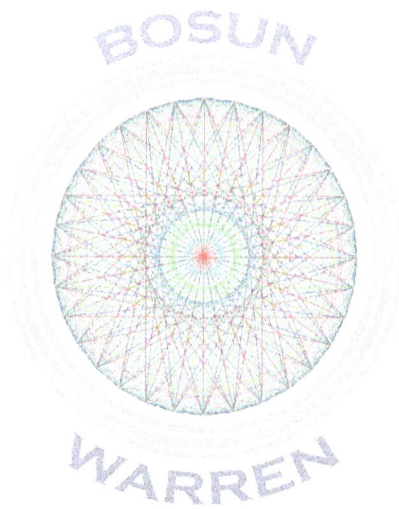

TABLE OF CONTENTS

Page

Boscawen-Ûn

by

James R Warren BSc MSc PhD PGCE

CHAPTER ZERO
PROLOGUE

There is an island in the North Atlantic Ocean where for more than four thousand years people have been trying to draw ovals.

It is easy to mock such strivers and the sentiments that inspire them. It is easy to mock as well as wicked and stupid. It is easy to dismiss the Wisdom of the Ancients as fatuous and primitive. But are we so wise?

What is Love? Is it the conjunction of Faith and Fellowship? Is true love always disinterested? One thing, and only one thing, we know for sure about the megalithic stone monument of Boscawen-Ûn: That it was not built by some single Cyclopean hero.

It was a structure, devoutly desired, and pulled together (literally and figuratively) by a congregation lost to history, "known only unto God", as they used to say of the obliterated dead.

Of course, the men and women who drew together the mighty stones of Boscawen-Ûn may have rationalised their efforts with some doctrine or an ideology. Or at least their scientists or magi may have done so for them. By the way, do not think that I include the females in order to simulate some fashionable imperative: I do so because if women did not handle the ropes of haulage in person (and they may have done so) then they encouraged and exhorted their men to do so. For it would have certainly crossed their minds that if the men were doing this, then they were not raping and pillaging their neighbors, or calling the same upon their kinfolk.

And so, an eternal congress gathered. A convention solemnly together as a silent chapter, forever, like the celebrants of some forgotten sacrament. A monument, hallowed through eons, a granite memory unforgettable, redolent of parents unknowable.

In the Breton province of North-Western France and throughout the British Isles to its North there are numerous "circles" and avenues of well-weathered stone pillars. Stonehenge is perhaps the most celebrated, but that and many others were thought to have been laid by anonymous persons some four or six thousand years before our days, for reasons and purposes unknown.

We have this approximate age because archaeologists and other scientists have used both carbon-dating and confirmatory dendrochronology upon bone and wood intimately associated with these foundations.

Indeed, wooden "circles" and other structures of pre-historic date have recently been detected by researchers. Usually, the timber has long rotted away and only postholes or something betray its former emplacement, though in some famous cases gnarled old remains have survived where the ground was or became propitious, for example in paludinal or littoral environments.

Some say that the megaliths of Britain were laid by extraterrestrial animals, or by slaves working under their direction. I do not despise such ideas, or dismiss them as impossible. But for my part I cannot think that alien visitors who had the technology to arrive at this planet would have been inclined to fool around with loops of rope.

As noted, the megalithic monuments, mostly of Neolithic or very early Bronze Age ("Copper Age") vintage, are concentrated especially in the Brythonic regions of western Britain and Brittany. Antiquaries and others would quickly remind us that the Brythonic Celts who came to endue these territories with their distinctive language and culture post-dated the stones by some two thousand years, placing those Celts in time somewhere halfway between the builders and ourselves.

By the way, the reason these districts of Europe, or "These Islands" as the Irish and some Scots prefer to style them are called Britain or Brittany, etcetera, is because the literate Greeks and Romans noted the devotion of the natives to Brigit or Bridget, otherwise called Bride or Bright or something. Brigit was a goddess

Boscawen-Ûn   Page 10 of 150   James R Warren of light and fire. The Romans came to call her Britannia, and named their part of the archipelago for her. There are "Bridestones" three hundred miles away in Cheshire and a Kirkbride four hundred miles North a few miles from my place of birth, for the Christians, when they came, wasted no time in adopting good St Bridget. This tutelary goddess appeared on British coins until into this century.

Perhaps our hoary standing stones at Boscawen-Ûn see themselves as competing with the goddess for permanence?

As we remarked, why these long-dead people built their great stone circuits we have no way of knowing. For some centuries, our own scholars thought that they were probable temples set to celebrate the lurid rites of barbarous peoples. Since around 1950AD, antiquarians and scientists have sometimes thought that the structures were more probably astronomical observatories or clocks, perhaps constructed to assist agriculture or other calendric activities, though multiple religious and technical uses were seldom ruled-out. Several have described or computed sophisticated astronomic alignments (appropriate to the years of construction) that seem inherent to the stones' layouts.

Notwithstanding, like the old English jurist, who always considered motive irrelevant or inscrutable, we shall not ask *why* the remote Ancients built their great stone structures, but *how*.

We will pause only to consider some of the mathematical "shortcomings" of the true circle which may or may not have perplexed the ancient builders. For sure, some of the vagaries of the numerical expression of geometric facts scandalised the scholars and magi of Europe's Classical past, especially the mathematical philosophers of the Pythagorean School who flourished in Ancient Greece.

It is long notorious that the true circle's ratio of its circumference to its diameter cannot be expressed in integral terms, or even as a rational fraction. This length ratio is Pi (π) or the Ludolphine Constant. It can be approximated to any degree but never calculated exactly. Its value is about 3.141592653589793238462643.... and π was "defined" in 1841 by Karl Weierstrass using the integral:-

$$\pi = \int_{-1}^{+1} \frac{dx}{\sqrt{1 - x^2}}$$

Equation 0.1

π is a Transcendental Number. Technically, this means that it is not the solution of any algebraic polynomial whose coefficients can be expressed as rational numbers, but in layman's terms it means that the digits after the decimal point not only never finish, but neither do they repeat in any sub-grouping.

So what is a Rational Number. It is simply any number that can be expressed as the ratio of two integers, such as $1/120 = 0.00833333333333333...$, or $1/6 = 0.1666666666666666666...$, or $1/7 = 0.142857142857143857...$, or $1/9 = 0.111111111111111111111...$

Note that sooner or later groups of rational number digits repeat in cycles.

This rather begs the question: What is an Irrational Number. Well the captious answer is that it is any number that is not rational, or in other words it is "incommensurable".

To find an Irrational Number take two lengths. A simple example is a unitary right-angle triangle adjacent with its perpendicular opposite, also equal to one. Their hypotenuse, according to Pythagoras and doubtless even earlier authorities is the square root of two or in mathematical language:-

$$\sqrt{2} \approx 1.4142135623730950488 \dots$$

Equation 0.2

The squiggly equals sign means that the long number is only *approximately* equal to the definition on the left hand side.

So the square root of two is irrational.

Why does any of this matter?

As I said, the non-integrality of certain numbers frequently offended our forebears.

So some modern scholars have suggested that the ancient monument builders systematically distorted their stone "circles" in order, Procrustes-like, to make them show simple integral relations between their lineaments. And did so without vitiating their utility.

This and kindred assertions are still most highly controversial, you might say heretical, throughout both of the worlds of modern science and of contemporary antiquarian scholarship. Many say that the pastoral hut-dwellers of our far millennia were entirely too stupid and too ignorant to do this sort of thing.

But we are on the side of the circle-builders and respect their ancient intelligence and their lost wisdom. We are heretics!

The Geometric Irrational Numbers and their Rational Estimators

When you draw simple shapes, whether with pencil and paper, a computer stylus or with ropes or rods laid over the ground you inevitably produce two or more lines that have a definite length ratio to one another.

The problem as the builders and planners of olden days saw it was that there was seldom a simple integer ratio or common integer divisor involved.

But the troublesome irrational could always be approximated by integer ratios, or simple arithmetic additions or multiplications of integer ratios. Needless to say, the quality of those rational approximations varied, but something adequate to the purpose in hand could usually be devised.

Architects and builders soon discovered that three key irrational ratios often arose:-

(1) π (Pi) or the Ludolphine Constant

This is the Ratio of the Circumference to the Diameter of a perfect circle as drawn with a pair of compasses or their scale equivalent and as we saw above has an unknown value better than 3.141592653589793238462643

(2)   The Square Root of Two, $2^{1/2}$

This is the diagonal of a unit square and is somewhere around 1.41421356237309503048

(3)   The (Greater) Ratio of Phidias, Φ

An outcome of folding and Fibonacci theories this quantity, otherwise known as the Golden Section or

Golden Ratio or something, was employed by the Ancient Greeks and later artists in the proportional planning of architecture and artwork.

Φ has an unknowable value adjacent to 1.618033988749894

The Ratio of Phidias is defined by:-

$$\Phi = \frac{1 + \sqrt{5}}{2} \approx 1.618033988749894$$

Equation 0.3

These are not of course the only irrational numbers to arise when you lay out a drawing, whether for a machine, a building, a megalithic henge or indeed for aesthetic pleasure.

There are theoretically an infinitude of ways in which any given irrational number can be approximated, but it is probably as convenient for us as it was for the Ancient Britons to use a simple addition and multiplication formula similar to this one:-

$$a = \frac{m}{n}\left(\frac{i}{j} + \frac{k}{l}\right)$$

Equation 0.4

where a is the Rational Approximation of the Irrational Number t = π, $2^{\frac{1}{2}}$, Φ, or whatever; and m, i , k are all Integers, whilst n, j and l are Positive Whole Numbers greater than zero.

It is always useful to compute and compare a measure of accuracy in the approximation and I recommend the Percentage Specific Defect, PSD, (do not confuse with the closely-related parameter, Population Standard Deviation, σ).

PSD is defined using:-

$$PSD = 100\left(\frac{t - a}{t}\right)$$

Equation 0.5

An advantage of PSD is that it is just about the simplest possible measure of relative displacement, and it preserves the sign of that displacement.

Except in cases of algebraic identity, PSD is almost never zero, but the smaller its absolute value the better.

An Example for Pi

Let us see how Equation 0.4 can be used to formalise the well-known approximation for π which is the rational fraction 22/7.

Define m = 1, n = 1, i = 22, j = 7, k = 0 and l = 1.
Then:-

$$a = \frac{m}{n}\left(\frac{i}{j} + \frac{k}{l}\right) = \frac{1}{1}\left(\frac{22}{7} + \frac{0}{1}\right) \approx 3.14285714285714$$

Equation 0.6

Certify that no denominator is zero or else you will get an overflow error because $1/0 = \infty$.

To assess the quality of our approximation we can use the PSD thus:-

$$PSD = 100\left(\frac{t - a}{t}\right)$$
$$\approx 100\left(\frac{3.14159265358979 - 3.14285714285714}{3.14159265358979}\right)$$
$$= -0.04024994347707$$

Equation 0.7

So the error of using 22/7 to approximate π is about four hundreds of a percentage point and since PSD is negative the approximation is a smidgen too big.

Table 0.1 presents the inputs and results for a brace of rational approximations for each of π, $2^{\frac{1}{2}}$, and Φ.

Cornwall

Britain is a very complex country, haunted for good and for evil by its complex past.

In the extreme south-west of Great Britain is the peninsular English county of Cornwall. Many claim it is not truly English because until two hundred years ago it had its own customs and language, very unlike those of the Anglo-Saxon East, but very similar to the mystic and communitarian, deeply-religious, tenor of the Welsh people or the Bretons, who equally disputably are French.

Like Wales and Brittany, Cornwall is humid, cool, windswept and lush, a land of bleak wastes and verdant valleys. A land redolent of peregrine saints, and of the idylls of the long-lost kings who patronised or persecuted their sacred missions. A land of wandering tourists, farmers and fishers, and of *appellations contrôlée*.

Cornwall is a land of tidal rias and sea-washed creeks, last resting places of rotting ships, now barely shadows in the sand. A land of cows, of unintelligible voices, of poets and painters, of sparkling subtropical light merging sky and sea, of turquoise waters and vernal dancers. Of empty granite engine houses, stronger than castles, and their long-smokeless rook-ridden chimneys. It is as mystics say, liminal.

The Enigmatic Monument of Boscawen-Ûn

The Cornish-language phrase "Boscawen-Ûn" means something like "Alder Tree Farm Pasture". Boscawen-Ûn is just off the A30 Hounslow (London) to Land's End road. It is about 5.5 miles or 8.85 Kilometers from Land's End. Boscawan-Ûn is 290 miles from London.

Analytic Value	Multiplier Numerator m	Multiplier Denominator n	Augend		Addend		Rational Estimate	Percentage Specific Defect
			Integral Denominator	Integral Numerator	Integral Denominator	Integral Numerator		
3.14159265358979	1	1	22	7	0	1	3.14285714285714	-0.040249943477707
3.14159265358979	1	1	355	113	0	1	3.14159292035398	-0.00000849136788
1.41421356237310	1	1	198	140	0	1	1.41428571428571	-0.0051019106886
1.41421356237310	1	2	99	70	140	99	1.41421356421356	-0.00000013014082
1.61803398874989	1	1	1	1	55	89	1.61797752808989	0.00348946069118
1.61803398874989	1	1	1	1	89	144	1.61805555555556	-0.00133290189271

Table 0.1
Six Rational Approximations of Familiar Irrational Numbers

Boscawen-Ûn is a vaguely oval array of nineteen short Early Bronze Age megaliths surrounding an eccentrically-placed long gnomon-like menhir which has a pronounced slope. The circuit stones are of granite except for one which is quartz. The lithology of the internal inclined stone is not known to me.

Amidst his orbit of attending ortholiths the lonely canting menhir bows his long apology to the eternal sovereignty of outraged Time. Is his dramatic lean through naked space an accident of ground creep and the vagaries of wind and weather? Or is he like the dying puissance of a god denied, tired of attending the ungrateful eons or the tailing story of a culture consummated? Or perhaps he is fatigued as old men tend to be while awaiting the mighty quoits of mightier hurlers who forgot to come to play?

When the menhir was set in place it may have been vertical. Among controversial modern theories is that the menhir is indeed a sun-shadow or moon-shadow gnomon arranged to indicate an astronomical or calendric event. This is considered along with the menhir being originally vertical; or originally at its present inclination. Third even more contentious interpretations involve the base of the inclined stone being in shadow or the top being in shadow only at certain fleeting conjunctions of celestial bodies, principally the Sun or the Moon. In the latter opinions, scholars sometimes allege that the Boscawen-Ûn apolith or other similar inclined stones elsewhere show signs of basal packing, as if the builders attempted to set a precise angle of inclination

Whatever; these score fellows dream on through the mists and the moonlight, the sunshine and the storm, careless of fleeting speculations.

The apolith was canting at some angle at least as early as 1749AD when the antiquarian William Stukeley surmised that it had been perturbed by treasure hunters.

According to GeoHack, Boscawen-Ûn is situated at 50° 5' 24.76" N, 5° 37' 10.49" W, or in terms of the British Ordnance Survey Grid Reference SW4113627470. The main "Boscawen-Ûn" Wikipedia article cites the NGR as SW412273 or SW 412274. The National Libraries of Scotland Side-by-Side applications utility gives 50° 5' 24" N, 5° 37' 10" W for the apolithic menhir or SW4121827357 as its NGR.

Figure 0.1 is an enhanced vertical aerial photograph of the monument taken from the NLS website. The eccentric menhir which may or may not be a gnomon is marked by the cross-pointer, for which the NLS co-ordinates pertain. The two white objects circled in yellow are sheep.

Figure 0.1
Aerial Plan View of the
Boscawen-Ûn Stone "Circle" Monument

By the way, and at the expense of being tedious, I am advised that the Cornish phoneme "Û" is pronounced "oo" somewhat as in the RP "cool" or "fool" (not "wool") and not the RP "bun". But in your accent and definitely mine I suppose the Cornish might forgive our manglings.

Figure 0.2
View of the Boscawen-Ûn Stone "Circle" Monument

CHAPTER ONE
THE ANCIENT ARTS OF TECTONIC GROUND PARTITION

This treatise concerns the planimetric geometry of a Cornish stone circuit as an expression of the planning and lay-out technologies of a remote past. Such technologies imply a knowledge of mathematical practice unappreciated until some sixty years ago.

The megalithic circuit at Boscawen-Ûn is one of some scores of ancient stone "circles" scattered throughout Great Britain. It is thought that the surviving monuments are but a fraction of those that existed. It is also thought that these figures of stone were installed between 6000BC and 2000BC in The Neolithic or Early Bronze Ages.

These megalithic stone "circles" are almost never circular in plan, deviating subtly, and I consider intentionally, from a perfect roundness.

No living man knows why any of these monuments were laid-out. The records of the Ancient Britons have not survived the decays of time, if those records ever existed. Archaeologists have provisionally determined that some of these megalithic circuits bear a relation to planetary or sidereal dispositions as such phenomena would have appeared at the time of construction.

If, as Bloody Mary claimed *Veritas Filia Temporis*, truth is the daughter of time, then Boscawen-Ûn belies the maxim. For the son of time is confoundment, and confusion the father of falsehood. Unlike some of our forebears, we cannot be sure whether the stones were laid by Satan in person; Or by helots under the hieratic authority of some nameless magus; Or by cheerful workmen whistling to the tune of their pragmatic foreman.

What we can do is compute conformations practicable using tools available to the era; Test the technical feasibility of plan realisations using experimental archaeology; Then use geophysical techniques to detect or deny the vestiges of postholes at predicted places.

It is known from Ancient Egyptian illustrations and scripts that that African nation used tensioned ropes as an adjunct to surveying and the laying-out of structures. Boscawen-Ûn is thought to antedate The Great Pyramids of Giza.

My thesis is that Ancient Britons used similar taut-rope techniques.

Taut Rope Technique

We may conjecture from common sense and slender evidence that primitive societies draw plans upon the ground using rope-and-stake methods that involved driving one or more firm stakes into the earth and tying a sturdy rope of hide or hemp to it. Some burly men would then be directed to pull the rope tight and orbit its further end around the stake.

The rope may or may not be suffered to slip. Slight sophistications of such a method would include an additional stake or stakes that might intercept the rope in its course and turn a circle into an ellipse or a more complex oval.

Those who read my 18 February 2013 essay about Roman marching camps in Britain, "Some Design Aspects of Roman Encampments in Britain: An Extended Study"[1.1] may recollect our elaborate though highly-provisional discussion of such methods.

If such were done then no example of such a rope or loop of rope has come down to us: Such organic material is readily re-used, or decays if discarded. More embarrassingly, I know of no surveyor's chain of iron or bronze that would lend itself to such employment and which has come to us from Antiquity.

The Ancient Egyptians[1.2,1.3] were very familiar with the use of calibrated ropes applied to the procedures of survey and build planning and illustrated such work in figurative murals. Such processes are literally geometry and lend themselves to imitation with paper, pin, thread and pencil and that shall assist our researches:- Or more precisely, our discussion of viable and efficient proposals.

The Greeks called the practitioners of calibrated-rope surveying harpedonaptai or "rope-stretchers". I suppose we could frame the adjective "harpedonaptal". But we should be careful to distinguish the use of open-strand taut ropes applied as simple radii from the use of closed loops of tensioned rope used to describe ellipses or ellipse-like closed curves or the arcs of elliptic curves. I shall take the liberty of specifying looped-rope planning as "koulography" from the Greek κουλούρα for a loop of material. It is an ugly word and you can doubtless improve upon it.

Simple and Multiple Figure Foci

We may provisionally classify the planar figures generable using ground-stakes according to the number of stakes set out. If one stake is driven we may classify the resulting figure (a circle or circular arc) as a MONOPOLAR MONOPOLE (forgive the pleonasm: it is merely to simply the definition of terms). Two stakes set apart would constitute the foci of an ellipse which is a DIPOLE. Four stakes would set the planar geometry of a TETRAPOLAR oval which may be a compound ellipse. Eight stakes an OCTOPOLE and so forth.

As we shall see, ground-figures may be inscribed as compounds of arcs and whilst superficially tetrapolar may actually be four discontinuous monopoles (i.e. four disjointed circular arcs).

The Character and Setting of Boscawen-Ûn

Figure 0.2 is a 7 August 2018 panoramic photograph of the monument taken from near ground level and at an azimuth approximating 105°. It is sourced from the Wikimedia Commons[1.4] article, and was contributed by Waterborough[1.5].

The phrase "Boscawen-Ûn" means something like "The meadow farm amidst the alders". It is an expression of the Cornish language.

The ancient oval of eighteen granite stones and a sole quartz example embowers a solitary gnomon of inclined granite set eccentrically in their midst. I say "gnomon" to lend you an image of its inclined sub-centrality, but its actual function, if any, and that of its circumjacent companions are lost to the history that did not yet exist as the annals of writing men.

Accordingly, I shall not speculate as to the design function of Boscawen-Ûn, for such guesses are vain. And what is the value of mystery if motive is discovered? Rather I shall propose a few likely MODII, or systematic methods (singular: MODUS), of megalith layout, the outcomes of the geometric imperatives of the specific technique chosen. Those methods may be arbitrary or pragmatic rather than scientific or ritualistic in intent.

The monument is set inconspicuously amidst scrub and gorse bushes about 500 meters South of the main A30 London to Land's End road about 314.4 miles from London; 5.4 from Penzance (Holy Headland) and 5.21 miles short of the end of the road at Land's End. Excepting isolated farmsteads, the nearest settlement is the hamlet of Crows-an-Wra (Witch's Cross) about one mile to the West along the A30.

Boscawen-Ûn is set amidst the moors and meadows of the granite promontory of Penwith at the extreme West end of Cornwall, itself the culminating peninsula of South-West England.

Penwith is Cornish for "the headland at the end". The word is cognate with the English "Land's End", the French "Finistère" and the Spanish "Finisterre" all of which refer to the windswept Atlantic termini of their respective Celtic extremities.

Penwith is a lovely if somewhat mystic pene-plateau of undulating granite set above obdurate granite cliffs. Many a fine ship, scudding home before the relentless wind, has fallen helplessly upon its wave-washed reefs. Whatever the circumstances, no visit to Penwith can ever be forgotten.

Penwith is a lonely land inhabited by prehistoric menhirs, disused tin and copper mines, ancient burial mounds, holy wellsprings and deeply-incised, lushly-wooded subtropical valleys that rush tinstone-laden waters to a foaming sea. In the center of the promontory the Ding-Dong Mine is said to have raised tin from The Bronze Age until 1915AD, a productive life of maybe four thousand years. Our Holy Savior is said to have wondered here as an adolescent when he visited with his uncle about the tin trade. Inevitably, ghosts, witches and Dark Age saints crowd out the mortal inhabitants even or perhaps especially at the height of the summer tourist season, but in a Winter's hurricane any castaway would find this a bleak landfall indeed.

Whether on the peaty windswept moors at Ding-Dong where one might turn from the lonely engine house to view a shimmering sky or sea, whether hearing the wave-tossed whistle that marks the fatal reef of The Runnel Stone, or adding your blooms to the votive flowers at Carn Euny or your rags to the Clouthy Tree or your naked body to the everlasting, the copious, the sacred (and very radioactive) waters of the baptistery ruined in the haunted woods of Madron, you will never forget any visit to Penwith. As the mystics say, it is liminal.

If you forgave my pleonasm, I also beg you to indulge my grosser repetitions: They are a licence of dotage.

The Boscawen-Ûn Monument: Its Place in Time

It is much easier to document the monument's place in space rather than in time despite the extreme mutability of the former.

Experts think that the structure was emplaced sometime during the Bronze Age in Britain, perhaps in the interval 3300 to 1200 BC. Therefore the "circle" is about four thousand years old. A twentieth circuit stone may have been removed, possibly during historic but undocumented events. When such a wide degree of uncertainty exists in the chronology, the imputation of intent on the part of its builders is even more hopeless than usual. 3300BC predates the great Pyramids of Old Kingdom Egypt by about eight hundred years, whilst 1200BC places us with Homer and the Trojan War. The twenty-one hundred years of the Bronze Age compares with that stretch of time between the Ptolemies and the Victorians. Now if we were told that a grand building with an Ionic pronaos was erected by Hellenistic Egyptians we would immediately assume (rightly or wrongly) that it was a temple, but if by Victorians a bank. Radiodating or other archaeological science may throw light upon the age of Boscawen-Ûn, but those and other chronological techniques raise problems of their own.

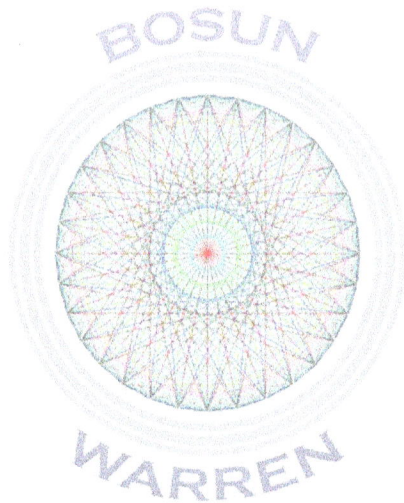

Data

According to Side-by-Side[2.1] the Northern (aerial) tip of the central Boscawen-Ûn menhir, the supposed "gnomon" is given by the co-ordinates imaged below:-

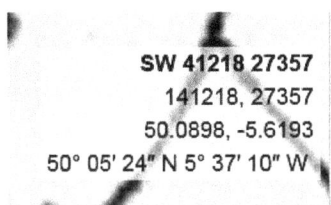

SW 41218 27357
141218, 27357
50.0898, -5.6193
50° 05' 24" N 5° 37' 10" W

The first coding, SW 41218 27357 is the UK Ordnance Survey Grid Reference and the second (141218, 27357) its National Grid Reference (NGR) numerical equivalent. The third co-ordinate is the Decimal Latitude and Longitude referenced to the Greenwich Observatory, London. The fourth set of figures is the Sexegesimal Latitude and Longitude in degrees, minutes and seconds of arc.

Figure 2.1 presents the Side-by-Side map-and-photograph pair from which I measured data:-

Figure 2.1
The National Library of Scotland Map-Photograph Pairing
that served as the
Basis of my Planimetry

Reproduced under a Creative Commons Attribution-NonCommercial-ShareAlike
4.0 International (CC-BY-NC-SA) licence
with the permission of the National Library of Scotland.

Note the very slight change of co-ordinates: An artefact of an unsteady hand.

Figure 2.2 marks the terrestrial Base of the "gnomon" menhir and its aerial Tip, as well as numbering the nineteen circuit stones in a counter-clockwise sense. Trespassing sheep, easily confused with standing stones, are marked with a red A and B.

You can see that the plan of the monument describes some kind of oblate oval, not necessarily a Ragazzo Oval[2.2] or a Cubic Superellipse[2.3], but somewhat suggestive of either.

Figure 2.1 was cropped and annotated to yield Figure 2.2:-

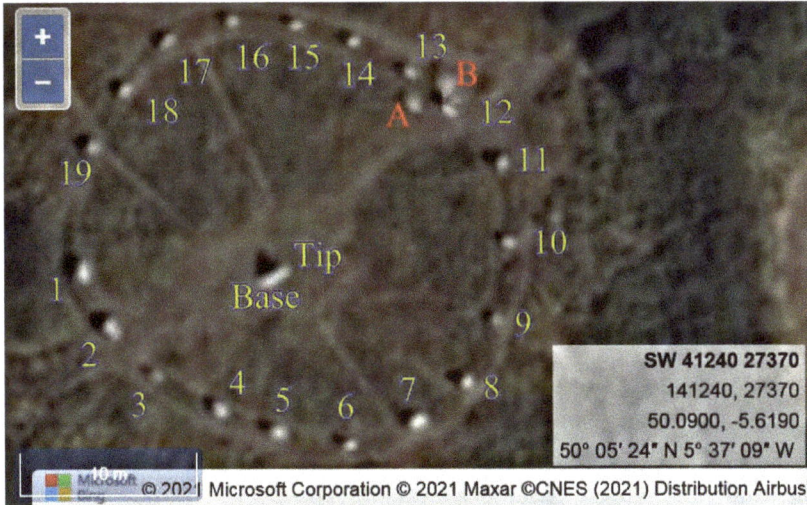

Figure 2.2
The Basis Photograph for Cartesian Position Measurements
Reproduced under a Creative Commons Attribution-NonCommercial-ShareAlike
4.0 International (CC-BY-NC-SA) licence
with the permission of the National Library of Scotland.

Semi-Manual Digitisation

I essayed two semi-automatic methods of digitising the planar Cartesian co-ordinates with my mouse:-

(A) *NLS Satellite Photograph*
       10 units to 10 meters

This involved direct noting of photograph co-ordinates from the website screen, as rendered to OS co-ordinates in the National Library of Scotland cartouche.

(B) *PhotoDraw Co-ordinates*
       35.5 mm to 10 meters

I copied the screen image into a JPEG file, and loaded the same into PhotoDraw® (Yes: I am easily conservative enough to retain and employ PhotoDraw).

Using the calibration of 35.5 PhotoDraw meters to 10 satellite photograph meters I computed the nineteen circuit co-ordinates as per the satellite image together with the positions of the apparent "gnomon" extremities and the instant positions of the two grazing sheep.

The nominal precision of the PhotoDraw co-ordinates was superior to that of the NLS co-ordinates.

The Resulting Raw Data in both sets is listed in Table 2.1:-

Zeroisation

From a strictly scientific viewpoint the rigid translation of the data co-ordinates to a chosen center is not necessary.

Notwithstanding that, the centering and later rotation of the co-ordinates greatly simplifies analysis without vitiating the mathematical integrity of the data.

Specialists have shown that any triangle has an infinite number of centers, and I suppose that by extension (arguing along Delaunay lines) the same applies to any Euclidean figure.

But I would also argue that some centers are more useful than others and accordingly I propose to center the plan upon the point that defines the (Unweighted) Planar Barycenter, tantamount to the mean of the x-coordinates; and the mean of the y-coordinates; (x_c, y_c) where:-

$$x_c = \frac{\sum_{i=1}^{n} x_i}{n}$$

Equation 2.1a

and:-

$$y_c = \frac{\sum_{i=1}^{n} y_i}{n}$$

Equation 2.1b

n is the Number of Circuit Stones, that is to say nineteen.

Scaling

Procedurally speaking, the PhotoDraw convention is to count vertical displacements from the top of the image and horizontal from the left. Accordingly, I INVERTED the *PhotoDraw y-data only* before moving forward.

For both the NLS and PhotoDraw series I then SCALED the data by the relevant Calibration Coefficient C_{cal} to reduce all co-ordinates to meters.

The net outcome of the two affine transformations of Translation and Scaling is specified by:-

$$x_i' = C_{cal}.(x_i - x_c)$$
Equation 2.2a

.

$$y_i' = C_{cal}.(y_i - y_c)$$
Equation 2.2b

These ordered transformations produce a Zeroised Scaled Inversion series (ZSI) expressed in meters.

Rotation

The outcome is a Rigidly-Rotated Scaled Zeroised Inverted (RRSZI) co-ordinate data series that expressed the land position of megalithic stone features relative to their planar barycenter. This RRSZI series is fit for plotting or other further processing.

Accordingly, Figure 2.3 displays the plan of the stones as the RRSZI series for both the NLS and PhotoDraw renditions.

On the whole, the conformation of the two plans is comparable, but the PhotoDraw rendition is more regular and about 2.5 meters wider all round.

I decided to continue further analyses using the PhotoDraw RRSZI series and it is this I present in Figure 2.4

With a little imagination and the eye of faith it is apparent that the stone circuit comprises four distinct sub-arcuate sectors. It is also very suggestive of the removal of a stone that would have been near or at the Western boundary. As we have remarked, such a removal may have occasioned at some remote unknown time.

I propose the sectoral scheme shown in Figure 2.5

If the profile is a little disrupted this may be due to "land creep" over the millennia. It may alternatively or additionally be due to intended or unintended evolutions of the chosen stone-deployment method (design MODUS).

The PhotoDraw series data upon which the scheme is based is presented in Table 2.2

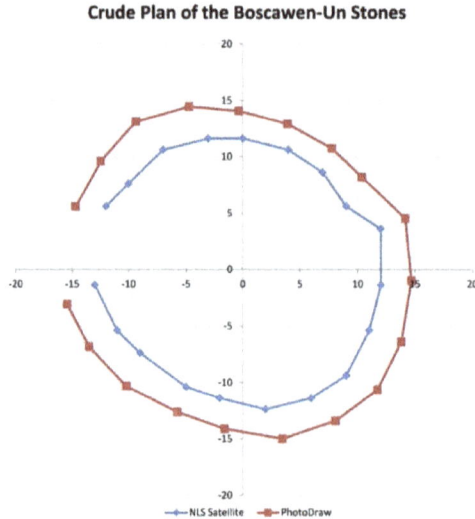

Figure 2.3
An EXCEL Plot of the ZSI Cartesian Co-ordinates of
Boscawen-Ûn Circuit Stone Features

		NLS Satellite Photograph 10 units to 10 meters		PhotoDraw Co-ordinates 35.5 mm to 10 meters	
	Scale	Easting	Northing		
1	Left	41199	27348	3.1	96
	Right	41209	27347	38.6	96
2	Left	41199	27350	3.1	87.9
	Right	41209	27350	38.6	87.9
Mean Dif.	Left	10		35.5	
	Right	10		35.5	

	x	y
Top Left	0	0
Top Right	162	0
Bottom Left	0	91
Bottom Right	162	91

Stone	Sheet	Easting	Northing	x	y	Inversion x	y
1	SW	41207	37359	16.1	52.8	16.1	38.2
2	SW	41209	27355	21.5	63.4	21.5	27.6
3	SW	41211	27353	31.1	22.8	31.1	68.2
4	SW	41215	27350	43.3	79.7	43.3	11.3
5	SW	41218	27349	55.2	83.9	55.2	7.1
6	SW	41222	27348	69.4	86.4	69.4	4.6
7	SW	41226	27349	82.4	81.9	82.4	9.1
8	SW	41229	27351	92.6	74.1	92.6	16.9
9	SW	41231	27355	98.5	62.2	98.5	28.8
10	SW	41232	27359	101	46.9	101	44.1
11	SW	41232	27364	99.5	31.4	99.5	59.6
12	SW	41229	27366	88.8	21.1	88.8	69.9
13	SW	41227	27369	81.5	13.9	81.5	77.1
14	SW	41224	27371	70.7	7.8	70.7	83.2
15	SW	41220	27372	58.7	4.5	58.7	86.5
16	SW	41217	27372	46.3	3.4	46.3	87.6
17	SW	41213	27371	33.3	7.2	33.3	83.8
18	SW	41210	27368	24.5	17.1	24.5	73.9
19	SW	41208	27366	18.2	28.4	18.2	62.6
Base	SW	41218	27357	52.7	54.5	52.7	36.5
Tip	SW	41219	27358	56.3	52.3	56.3	38.7
A	SW	41227	27367	82.1	20.4	82.1	70.6
B	SW	41229	27368	88.8	17	88.8	74

Table 2.1
The Raw NLS and PhotoDraw Feature Co-ordinate Data

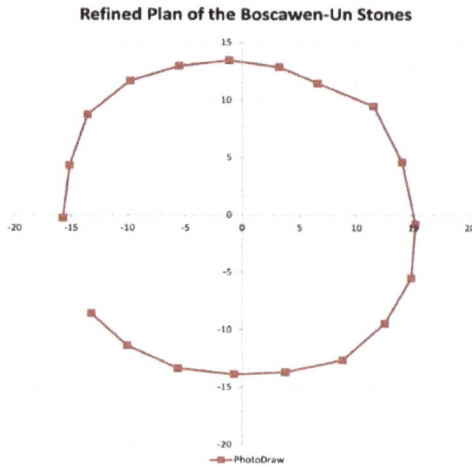

Refined Plan of the Boscawen-Un Stones

Figure 2.4
An EXCEL Plot of the RRSZI Cartesian Co-ordinates of
Boscawen-Ûn Circuit Stone Features

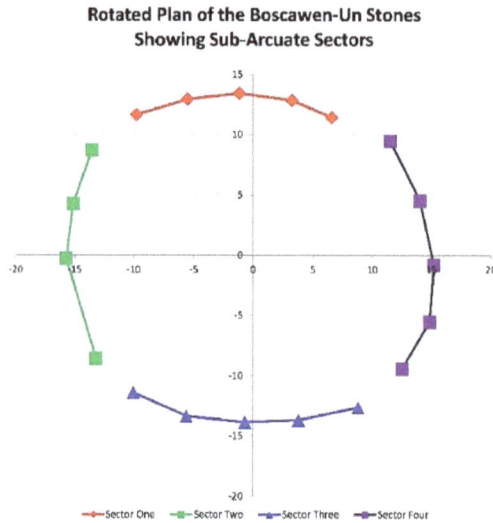

Rotated Plan of the Boscawen-Un Stones
Showing Sub-Arcuate Sectors

Figure 2.5
An EXCEL Plot of the RRSZI Cartesian Co-ordinates of
Boscawen-Ûn Circuit Stone Features with
Suggested Sub-Arcuate Sectors

PhotoDraw Co-ordinates
35.5 mm to 10 meters

Factor: 0.355

3.1	96
38.6	96
3.1	87.9
38.6	87.9
35.5	
35.5	

	x	y
Top Left	0	0
Top Right	162	0
Bottom Left	0	91
Bottom Right	162	91

Mean x 21.158
Mean y 16.62334
Slope Angle -0.380481777
Slope Angle° -21.8

Inversion				Scaled Inversion (meters)		Zeroised Scaled Inversion (meters)		By-Eye Rigidly Rotated Scaled Zeroised Inversion (meters)		Hand Plot Rounded RRSZI (meters)	
x	y	x		x	y	x	y	x	y	x	y
16.1	52.8	16.1	38.2	5.7155	13.561	-15.4425	-3.06234	-13.2009	-8.57819	-13.2	-8.6
21.5	63.4	21.5	27.6	7.6325	9.798	-13.5255	-6.82534	-10.0235	-11.3602	-10	-11.4
30.9	73.2	30.9	17.8	10.9695	6.319	-10.1885	-10.3043	-5.63318	-13.3511	-5.6	-13.4
43.3	79.7	43.3	11.3	15.3715	4.0115	-5.7865	-12.6118	-0.68905	-13.8588	-0.7	-13.9
55.2	83.9	55.2	7.1	19.596	2.5205	-1.562	-14.1028	3.787047	-13.6744	3.8	-13.7
69.4	86.4	69.4	4.6	24.637	1.633	3.479	-14.9903	8.797133	-12.6263	8.8	-12.6
82.4	81.9	82.4	9.1	29.252	3.2305	8.094	-13.3928	12.48884	-9.42921	12.5	-9.4
92.6	74.1	92.6	16.9	32.873	5.9995	11.715	-10.6238	14.82256	-5.51351	14.8	-5.5
98.5	62.2	98.5	28.8	34.9675	10.224	13.8095	-6.39934	15.19843	-0.81329	15.2	-0.8
101	46.9	101	44.1	35.855	15.6555	14.697	-0.96784	14.00538	4.559365	14	4.6
99.5	31.4	99.5	59.6	35.3225	21.158	14.1845	4.534658	11.46751	9.470605	11.5	9.5
88.8	21.1	88.8	69.9	31.524	24.8145	10.366	8.191158	6.582752	11.45497	6.6	11.5
81.5	13.9	81.5	77.1	28.9325	27.3705	7.7745	10.74716	3.227364	12.86578	3.2	12.9
70.7	7.8	70.7	83.2	25.0985	29.536	3.9405	12.91266	-1.13665	13.45259	-1.1	13.5
58.7	4.5	58.7	86.5	20.8385	30.7075	-0.3195	14.08416	-5.52705	12.95829	-5.5	13
46.3	3.4	46.3	87.6	16.4365	31.098	-4.7215	14.47466	-9.75927	11.6861	-9.8	11.7
33.3	7.2	33.3	83.8	11.8215	29.749	-9.3365	13.12566	-13.5433	8.719712	-13.5	8.7
24.5	17.1	24.5	73.9	8.6975	26.2345	-12.4605	9.611158	-15.1387	4.296395	-15.1	4.3
18.2	28.4	18.2	62.6	6.461	22.223	-14.697	5.599658	-15.7255	-0.25879	-15.7	-0.3
52.7	54.5	52.7	36.5	18.7085	12.9575	-2.4495	-3.66584	-0.91295	-4.31335	-0.9	-4.3
56.3	52.3	56.3	38.7	19.9865	13.7385	-1.1715	-2.88484	-0.01638	-3.11359	0	-3.1
82.1	20.4	82.1	70.6	29.1455	25.063	7.9875	8.439658	4.282063	10.8024	4.3	10.8
88.8	17	88.8	74	31.524	26.27	10.366	9.646658	6.042226	12.80638	6	12.8

Max 14.697 14.47466 Max 15.19843 13.45259 Max 15.2 13.5 Max
Min -15.4425 -14.9903 Min -15.7255 -13.8588 Min -15.7 -13.9 Min
Axis Length 30.1395 29.465 Axis Length 30.92392 27.31143 Axis Length 30.9 27.4 Axis Length
15.06975 Mean Length, a 15.46196 Mean Length, a 15.45 Mean Length, a
14.7325 Mean Length, b 13.65572 Mean Length, b 13.7 Mean Length, b

Table 2.2
The Raw and Refined Co-Ordinate Data for
The Boscawen-Ûn Circuit Stones
(PhotoDraw mensuration scheme)

The Figure Perimeter Circumference, P

A key parameter of stone circuit morphology is the Figure Perimeter Circumference length P.

Without prejudice of model fitment we may consider P to be comprised of straight line segments between successive stones. This will inevitably underestimate the trace of any viable koulographic modus.

P is defined by:-

$$P = \sum_{i=1}^{n-1} p_i = \sum_{i=1}^{n-1} \sqrt{(u_{i+1} - u_i)^2 + (v_{i+1} - v_i)^2}$$

Equation 2.4

The Vertex Radius, ρ_i

The Vertex Radius ρ_i is the straight-line distance between the assumed Figure Center (x_c, y_c) and the circumferential point at (u_i, v_i) as represented by one of the circuit stones.

ρ_i is defined by:-

$$\rho_i = \sqrt{u_i^2 + v_i^2}$$

Equation 2.5

For centered data.

The Raw Vertex Angle, θ_i

The Raw Vertex Angle is the radial angle of a stone setting swept counter-clockwise (sinistrally) from the Figure Centre with respect to zero at the Abscissal East.

The Raw Vertex Angle, θ_i, is given by:-

$$\theta_i = \pi + \text{atan2}(u_i, v_i)$$

Equation 2.6

The Ludolphine Constant simplifies further arithmetic developments by ensuring that θ_i is always positive and monotonically-increasing.

The Corrected Vertex Angle, θ'_i

The Corrected Vertex Angle is derived from θ_i and is a conditional variable defined by:-

$$\theta'_i \begin{cases} \theta_i < \theta_1 & (2\pi + \theta_i) - \pi \\ \theta_i \geq \theta_1 & \theta_i - \pi \end{cases}$$
Equation 2.7

The Sectoral Angle, ψ_i

The Sectoral Angle, ψ_i, is the angle between two Vertex Radii to adjacent circuit stones.
It is given by:-

$$\psi_i = \theta'_i - \theta'_i$$
Equation 2.8

The Heronian Half-Perimeter, s_i

The Heronian Half-Perimeter is a preliminary variable in the calculation of the Area of a Sectoral Triangle from three known sides by means of Heron's Formula.
s_i is defined by:-

$$s_i = \frac{p_i \rho_i \rho_{i+1}}{2}$$
Equation 2.9

.

The Heronian Sector Area, η_i

The Heronian Sector Area is given by:-

$$\eta_i = \sqrt{s_i(s_i - p_i)(s_i - \rho_i)(s_i - \rho_{i+1})}$$
Equation 2.10

The Included Angle Sector Area, α_i

The Included Angle Sector Area is given by:-

$$\alpha_i = \frac{\rho_i \rho_{i+1} . \sin \psi_i}{2}$$
Equation 2.11

The Figure Area, A

The Area of the Figure, A, is the area covered by the stone circuit given that it is a Euclidean plane and that its boundary is demarcated by straight-line segments from stone to stone.
It may be expressed as:-

$$A = \sum_{i=1}^{n-1} \alpha_i = \sum_{i=1}^{n-1} \eta_i = \sum_{i=1}^{n-1} \frac{\rho_i \rho_{i+1} . \sin \psi_i}{2}$$
$$= \sum_{i=1}^{n-1} \sqrt{s_i(s_i - p_i)(s_i - \rho_i)(s_i - \rho_{i+1})}$$
Equation 2.12

The Simple Nucleation Factor, N_z

The Simple Nucleation Factor, N_z, is a basic measure of the roundness and compactness of a distribution, such as a "circle" of stones which are locatable in terms of their radial distances from a common center.
N_z has the dimensions L^1, but it can be rendered to a dimensionless form by dividing by the Mean Point Radius from the

Center, ρ_μ, [in our context the center is (x_c, y_c)]. This yields the Corrected Nucleation Factor, N_c.

Algebraically:-

$$\rho_\mu = \frac{\sum_{i=1}^n \rho_i}{n}$$

Equation 2.13

$$N_z = \frac{A}{P}$$

Equation 2.14

$$N_c = \frac{N_z}{\rho_\mu}$$

Equation 2.15

The Polar Profile

The Polar Profile is a representation of the vertex (stone) conformation in the angle-radius domain. In practice it is a plot of the n vertex points (stone positions) in the domain (θ'_i, ρ_i) as is shown below.

It is self-evident that the angle-radius relationship can itself be subjected to statistical analysis as well as facilitating a graphical comparison with other geometrical forms.

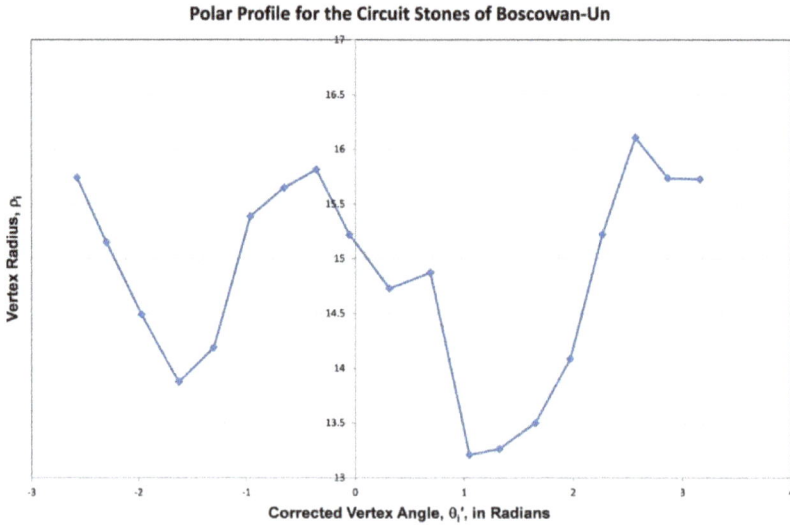

Polar Profile for the Circuit Stones of Boscowan-Un

Figure 2.5
A Polar Profile of the Stone Circuit of Boscawen-Ûn

Mean Squared Error, ε_μ

When there is a second deployment of points, for example those of another megalithic circuit monument or a theoretical closed shape whose perimeter is computed at suitable intervals; then we may define the difference between the two shapes in terms of the difference between their respective points. Clearly, it is helpful if both vertex sets are sampled at the same angular displacements, and that the size n (the object count) of the sets is the same.

Mean Squared Error (MSE), ε_μ, is defined as:-

$$\varepsilon_\mu = \frac{\sum_{i=1}^{n}(\rho_i - r_i)^2}{n}$$
Equation 2.16

This metric is very closely allied to similar statistics such as Root Mean Square Error (RMSE); (Population) Standard Deviation; (Percentage) Specific Defect and kindred descriptive statistics any of which may of course be preferred in a context.

As it stands, the dimensionality of ε_μ is L^2 and mathematically it is therefore an area dependent for its scale upon the units employed. To reduce ε_μ to a dimensionless statistic we could divide by a suitable area and I propose the sum of object Vertex Radii Squared as a suitable and convenient option. Therefore Dimensionless MSE ε'_μ may be defined as:-

$$\varepsilon'_\mu = \frac{\sum_{i=1}^{n}(\rho_i - r_i)^2}{n} \div \frac{\sum_{i=1}^{n}(\rho_i)^2}{n} = \frac{\sum_{i=1}^{n}(\rho_i - r_i)^2}{\sum_{i=1}^{n}(\rho_i)^2}$$

Equation 2.17

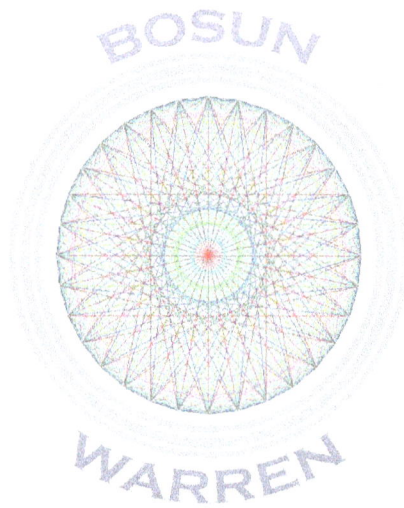

CHAPTER THREE
PRINCIPLES OF MODUS TREATMENT

The late genius Umberto Eco had Brother Jorge say that he would leave a certain knotty problem to "younger men".

I really understand this.

I do it all the time.

There is a certain phenomenon long observed in certain sandstones and slates called the "reduction spot".

It is a roughly circular or elliptical light green spot maculating an otherwise reddish rock.

Structural geologists and other experts in rock fabric have long recognised these spots to be sections through originally-spherical zones of nucleated bleaching, and that those marks that are elliptical were originally circular sections and hence may be used to quantify the crushing or stretching strain the rock body suffered through its history.

Implicit to the treatment is that the volume of the discolored inclusion is conserved upon deformation and that therefore the area of the elliptic section is identical to that of its original circle.

The diligent microscopist frequently observed tiny crystals of zircon at the center of these reduction ellipsoids: For as we almost said these spots were invariably sections through triaxial ellipsoids.

Now zircon includes the radioactive metal zirconium which is 2.8 weight percent Zr^{96}, an isotope that emits beta particles, essentially travelling electrons with a finite path length in rock. (Note that 49.76624% of a Zircon is Zirconium: The rest is either Silicon or Oxygen).

For fifty years I have surmised that it was these beta electrons which reduced red Fe^{3+} ferric ions to green Fe^{2+} ferrous ions.

You may be tempted to borrow the words of another of Eco's characters of fiction (Abbot Abo of Fossanova) to question the relevancy of this obscure disquisition.

The point is that once the triaxial ellipsoid of chemical reduction had reached its maturity in the indurated sedimentary deposit its volume was conserved forever.

All or part of this thinking might be quite wrong, or more likely oversimplified.

But a Theory of the Conservation of Area is a reasonable starting-point for our modelling of the genesis of the Boscawen-Ûn monument, whilst recognising that "land-creep" and the vicissitudes of eons have distorted the circuit.

Working Principles

We are in a position provisionally to enunciate a small number of principles about the Boscawen-Ûn megalithic circuit (we eschew talk of "stone circles" as highly-prejudicial) and the degree to which the *trace* of the circuit approximates ideal figures.

We define the trace as the more or less perfect and continuous geometric plan drawn by a loop of rope dragged over the ground, and the putative mark a dragged perpendicular pole would make thereupon.

Now having read thus far you have undoubtedly ascertained that we can say nothing about the design trace essayed by the authors of the monument: Whatever they wrote of their plans, if anything, has not survived the intervening millennia.

Therefore we are reduced to assumptions:-

(1)   The existing disposition of the circuit stones may be approximated by one or more ellipses, including one or more circles.

(2)   The actual best-fit figure might not be elliptical.

(3)   The measured stone circuit that exists in 2021AD is an arrangement of stones that may successively be connected by a continuous envelope of straight-line *segments*.

(4)   This *segmental* envelope has a calculable area that is *by definition* the area of the fitted curvilinear ideal figure.

(5)   The perimeter of the fitted figure is roughly the sum of the inter-stone straight-line segments.

(6) The Ratio of the Area of the Design Trace to its Perimeter, is about the same as the Area of the known segmental envelope to its Perimeter.

(7) The fitted ideal figure should minimise the Mean Squared Error (MSE) with respect to the measured stone circuit.

(8) In particular, because the intended parameters of the design circuit cannot be known; therefore only the Radius Displacements of the Ideal and Measured stone positions with respect to a Common Center can be determinants of Mean Standard Error.

Clearly, not all of these criteria are independent.

The Measured and Computed Dimensions of the Boscawen-Ûn Stone Circuit

The Mean Length of the Longer Semi-Axis of the Boscawen-Ûn Circuit, a is 15.46196194 meters and the Mean Length of the Shorter Semi-Axis, b, is 13.65571571 meters.
Algebraically:-

$$\alpha = ab = 15.46196194 \times 13.65571571 = 211.1441566$$
Equation 3.1

$$\beta = \frac{a}{b} = \frac{15.46196194}{13.65571571} = 1.132270345$$
Equation 3.2

Accordingly the product ab is 211.1441566 square meters and the dividend a/b is 1.132270345 dimensionless, given measurement upon the RRSZI basis.

The Mean Vertex (Point) Radius ρ_μ, is 14.84128522 meters.

In such terms the Area of the Segmental Envelope is 615.1207951 square meters and the Segmental Perimeter is

85.05769239 meters, as computed upon my EXCEL® Figure of Plan, from which the Simple Nucleation Factor, N_Z is:-

$$N_z = \frac{A}{P} = 7.231806764 \ meters$$

and the Corrected Nucleation Factor, N_C, is:-

$$N_c = \frac{N_z}{\rho\mu} = 0.487276314$$

The Area and Perimeter of a Circle

A Circle is a special case of an Ellipse whose major and minor (semi)-axes are equal (i.e. a=b).
The Area of a Circle, A_γ, is:-

$$A_\gamma = \pi r^2$$
Equation 3.3

and its Perimeter, P_γ, is:-

$$P_\gamma = 2\pi r$$
Equation 3.4

Therefore a Circle's Simple Nucleation Factor, N_z, is:-

$$N_z = \frac{A_\gamma}{P_\gamma} = \frac{\pi r^2}{2\pi r} = \frac{r}{2}$$
Equation 3.5

In this context, r is Radius.
The radius of a Circle that has the same sequential envelope Area, A, as Boscawen-Ûn is given by:-

$$r_\gamma = \sqrt{\frac{A}{\pi}} = \sqrt{\frac{\pi ab}{\pi}} = \sqrt{\frac{615.1207951}{3.141592654}} = \sqrt{ab} = 13.99282067$$

Equation 3.6

The Area of the General Ellipse

The Area of an Ellipse is given exactly by:-

$$A_\varsigma = \pi ab$$
Equation 3.7

Therefore, were Boscawen-Ûn an ellipse which it is not, its Area would be 663.3289313 square meters.

The Perimeter of the General Ellipse[3.1]

The Perimeter of the Ellipse, P_ς, cannot be computed exactly. It is analytically defined by:-

$$P_\varsigma = 4a \int_0^{\frac{\pi}{2}} \sqrt{1 - e^2.(\sin\theta)^2}\,.d\theta = 4a.E(e)P_\varsigma$$

$$= 4a \int_0^{\frac{\pi}{2}} \sqrt{1 - e^2.(\sin\theta)^2}\,.d\theta = 4a.E(e)$$
Equation 3.8

where:-

$$e = \sqrt{1 - \frac{b^2}{a^2}}$$
Equation 3.9

and E(e) is a Complete Elliptic Integral of the Second Kind:-

$$E(e) = \int_{0}^{\frac{\pi}{2}} \sqrt{1 - e^2.(\sin\theta)^2}.d\theta$$
Equation 3.10

This Ellipse Perimeter P_ς may also be approached discretely using the Ivory-Bessel Summation:-

$$P_\varsigma = \pi(a + b)\left[1 + \sum_{n=1}^{\infty}\left(\frac{(2n-1)!!}{2^n n!}\right)^2 \cdot \frac{h^n}{(2n-1)^2}\right]$$
Equation 3.11

where the auxiliary parameter h is given by:-

$$h = \frac{(a-b)^2}{(a+b)^2}$$
Equation 3.12

All of these classical forms are, however, too expensive for our necessarily approximate purposes and accordingly I recommend the use of the Ramanujan II approximation:-

$$P_\varsigma \approx \pi(a + b)\left[1 + \frac{3h}{10 + \sqrt{4 - 3h}}\right]$$
Equation 3.13

Therefore, we may express the Ellipse Simple Nucleation Factor, N_ζ, as:-

$$N_\varsigma = \frac{A_\varsigma}{P_\varsigma} = \frac{\pi a b}{\pi(a+b)\left[1 + \dfrac{3h}{10 + \sqrt{4 - 3h}}\right]}$$

$$= \frac{\pi a b}{\pi(a+b)\left[1 + \dfrac{3\dfrac{(a-b)^2}{(a+b)^2}}{10 + \sqrt{4 - 3\dfrac{(a-b)^2}{(a+b)^2}}}\right]}$$

Equation 3.14

which reduces to:-

$$N_\varsigma = \frac{A_\varsigma}{P_\varsigma} = \frac{a b}{(a+b)\left[1 + \dfrac{3h}{10 + \sqrt{4 - 3h}}\right]}$$

$$= \frac{a b}{a+b} \cdot \left(\frac{1}{1 + \dfrac{3h}{10 + \sqrt{4 - 3h}}}\right)$$

Equation 3.15

I have neglected the approximately-equal sign because for the given data Ramanujan II is as good as the analytic perimeter to fifteen decimal places.

The Mean Standard Error and Plot Comparisons

I need hardly remind readers that neither a circle nor a simple ellipse credibly represent the Boscawen-Ûn megalithic circuit, or throw any credible light upon its planning or construction.

Notwithstanding that, and for tutorial purposes only, we will plot a circle and an ellipse centered upon the barycentric origin each of which have the same area as the *segmental* circuit.

The needed Circle Radius r_ς is as given by Equation 3.6 (i.e. 13.99282067 meters).

The needed Ellipse Semi-Axes a and b may be estimated using:-

Determination of the Ellipse (Semi) Axis Values

Ellipse axes can be roughly estimated from a drawing or of course the ground, but we will attempt to employ a more scientific approach amenable to computation.

Apollonius Theorem

Apollonius Theorem specifies the Ellipse Minor Semi-Axis b and the Ellipse Major Semi-Axis a in terms of the Auxiliary Parameters M and N defined as below:-

$$M = a^2 + b^2$$
Equation 3.16

$$N = ab$$
Equation 3.17

From which we may compute a and b as:-

$$a = \frac{1}{2}\left(\sqrt{M + 2N} + \sqrt{M - 2N}\right)$$
Equation 3.18

$$b = \frac{1}{2}\left(\sqrt{M + 2N} - \sqrt{M - 2N}\right)$$
Equation 3.19

The Bounds of the Ellipse Circumference (Perimeter), P_ς

The Bounds of the Ellipse Perimeter conform to the range:-

$$4(a^2 + b^2) \leq P_\varsigma \leq \sqrt{2}.\pi.(a^2 + b^2)$$
Equation 3.20

Accordingly, it is possible for us to write:-

$$P \approx \sqrt{(4(a^2 + b^2)).\left(\sqrt{2}.\pi.(a^2 + b^2)\right)}$$
$$\approx \sqrt{\left(2^{2.5}.\pi.(a^2 + b^2)\right)}$$
$$\approx \sqrt{(2^{2.5}.\pi.M)}$$
Equation 3.21

where P is the Actual Perimeter of the Boscawen-Ûn Segmental Envelope as surveyed.

from Equation 3.21 it follows that:-

$$M \approx \frac{P^2}{\pi.(2^{1.25})^2} \approx \frac{P^2}{\pi.2^{2.5}}$$
Equation 3.22

Noting the constant:-

$$\frac{1}{\pi.2^{2.5}} \approx 0.056269769759819$$

We may further approximate M as:-

$$M \approx 0.056269769759819P^2$$
Equation 3.23

Reminding ourselves that A is the Actual Area of the Boscawen-Ûn Segmental Envelope as computed from survey data we may approximate N in these terms:-.

$$N = ab \approx \frac{A}{\pi}$$
Equation 3.24

From these facts we compute that our equivalent ellipse, by definition of the same area as the Boscawen-Ûn envelope, has Major Semi-Axis a = 16.0993315367745 and Minor Semi-Axis b = 12.1619355058334. Therefore, the "fitted" ellipse has an area of 615.1208. The Percentage Standard Defect PSD(A,A$_ç$) =

0.000000000000018, the small discrepancy being due to MathCad Express computational error.

The Eccentricity, e, and the Linear Eccentricity, c

The Eccentricity of the Ellipse is given by:-

$$e_\varsigma = \frac{c_\varsigma}{a_\varsigma} = \sqrt{1 - \left(\frac{b_\varsigma}{a_\varsigma}\right)^2}$$

Equation 3.25

and the length of the Linear Eccentricity, c_ς, by:-

$$c_\varsigma = \sqrt{a_\varsigma^2 - b_\varsigma^2}$$

Equation 3.26

For the relevant ellipse e_ς = 0.655228108666097 and c_ς = 10.5487345536292 (meters).

Crucially for our interpretations c_ς is the x-axis distance from the barycentric origin (the "center") to either of the ellipse foci.

The Outcome Perimeter

The Outcome Perimeter P_ς is given by:-

$$P_\varsigma = 4a_\varsigma \int_0^{\frac{\pi}{2}\pi/2} \sqrt{1 - \left(1 - \frac{b_\varsigma^2}{a_\varsigma^2}\right)(\sin\theta)^2} \, . \, d\theta$$

$$= 89.2167545828461$$

Equation 3.27

PSD(P,P_ς) is -4.88969554191082, showing that the reconstructed perimeter length is within five percent of the actual Boscawen-Ûn segmental perimeter, an accuracy which is about as good as can be expected from "fitting" a single ellipse.

<u>Tabulations and Plots</u>

Table 3.1 is a quotation of part of the Figure of Plan as it applies to further graphical developments.

Further to that I assembled computed specifiers of the plottable Cartesian points for the Boscawen-Ûn stone positions; the equivalent-area Circle centered upon the barycenter; and the equivalent-area Ellipse centered upon thee barycenter.

These values are presented in Table 3.2

Figure 3.1 is an EXCEL® Plot that compares the original Boscawen-Ûn plan data with the resulting Circle and Ellipse models. Clearly, the three co-radial points should be co-linear with each other and with the (barycenter) origin: Any apparent departure from co-linearity is due to my poor scaling of the EXCEL plot, a task performed by hand and eye.

A	615.1208
P	85.057692
N_r	7.2318068
N_c	0.4872763

COMPUTED

Stone Serial Number	Perimeter Segment p_i	Vertex Radius ρ_i	Corrected Vertex Angle (degrees)	Corrected Vertex Angle (radians)	Raw Vertex Angle (radians)	Sectoral Angle (degrees)	Sectoral Angle (radians)	Heronian Half Perimeter s_i	Included Angle Sector Area α_i	Heronian Sector Area η_i
1	4.2231573	15.743213	-146.9834	-2.565345	0.5762481	15.560195	0.2715766	17.558217	31.990319	31.990319
2	4.8206856	15.150064	-131.4232	-2.293768	0.8478247	18.547102	0.323708	17.230804	34.915691	34.915691
3	4.9701268	14.490859	-112.8761	-1.97006	1.1715327	20.029763	0.3495853	16.668471	34.434839	34.434839
4	4.4798975	13.875956	-92.84636	-1.620475	1.521118	18.326163	0.3198519	16.272467	30.953199	30.953199
5	5.1185288	14.18908	-74.5202	-1.300623	1.8409699	19.386246	0.3383538	17.348182	36.239351	36.239351
6	4.8836699	15.388756	-55.13395	-0.962269	2.1793237	18.080729	0.3155682	17.960551	37.369066	37.369066
7	4.5583991	15.648676	-37.05322	-0.646701	2.494892	16.649654	0.2905913	18.010927	35.453884	35.453884
8	4.7152233	15.814779	-20.40357	-0.356109	2.7854832	17.340492	0.3026487	17.875091	35.870827	35.870827
9	5.5035305	15.22018	-3.063074	-0.053461	3.0881319	21.095413	0.3681844	17.726272	40.342858	40.342858
10	5.5282061	14.728833	18.032339	0.3147237	3.4563163	21.519737	0.3755903	17.564852	40.177433	40.177433
11	5.2724372	14.872666	39.552076	0.6903139	3.8319066	20.563527	0.3589012	16.678402	34.508696	34.508696
12	3.6399187	13.2117	60.115602	1.0492152	4.1908078	15.802399	0.2758039	15.058008	23.861441	23.861441
13	4.4032881	13.264398	75.918002	1.3250191	4.4666117	18.911608	0.3300698	15.584107	29.020142	29.020142
14	4.4181458	13.500529	94.82961	1.6550889	4.7966816	18.269926	0.3188704	16.003228	29.812109	29.812109
15	4.4192866	14.087781	113.09954	1.9739593	5.1155519	16.766313	0.2926274	16.86616	30.936849	30.936849
16	4.8081208	15.225252	129.86585	2.2665866	5.4081793	17.35903	0.3029722	18.070461	36.584925	36.584925
17	4.7022427	16.107549	147.22488	2.5695589	5.7111515	16.93104	0.2955024	18.273161	36.908842	36.908842
18	4.5928275	15.736531	164.15592	2.8650613	6.0066539	16.786897	0.2929866	18.028489	35.740325	35.740325
19		15.727618	180.94282	3.1580479	0.0164552					
Total	85.057692	266.2568				327.92623	5.7233925	308.77785	615.1208	615.1208
Mean	4.7254274	14.841285				18.218124	0.3179662	17.154325	34.173378	34.173378
Pop.St.Dev	0.4492838	0.8970934				1.7099499	0.0298443	0.9119323	3.9938933	3.9938933
						32.073766	0.5597928			
						360	6.2831853			

Stone Serial Number	MEASURED EX PLAN Perimeter Segment p_i	MEASURED EX PLAN Vertex Radius ρ_i	MEASURED EX PLAN Sectoral Angle (degrees)	DIFFERENCES Perimeter Segment p_i	DIFFERENCES Vertex Radius ρ_i	DIFFERENCES Sectoral Angle (degrees)	Included Angle Sector Area α_i	Heronian Sector Area η_i	COMPUTED Heronian Sector Area η_i	POLAR PROFILE Corrected Vertex Angle (radians)	POLAR PROFILE Vertex Radius ρ_i
1	4.3	15.7	15.8	-0.076843	0.0432127	-0.239805				-2.565345	15.743213
2	4.82	15.2	19	0.0006856	-0.049936	-0.452898				-2.293768	15.150064
3	4.92	14.52	19.3	0.0501268	-0.029141	0.729763				-1.97006	14.490859
4	4.5	13.92	17.9	-0.020103	-0.044044	0.4261628				-1.620475	13.875956
5	5.08	14.3	19.8	0.0385288	-0.11092	-0.413754	37.2984	36.2814	36.23935103	-1.300623	14.18908
6	4.91	15.4	18	-0.02633	-0.011244	0.0807288				-0.962269	15.388756
7	4.55	15.62	16.8	0.0083991	0.0286758	-0.150346				-0.646701	15.648676
8	4.7	15.74	17	0.0152233	0.0747794	0.3404923	34.9747	31.8041	35.87082739	-0.356109	15.814779
9	5.56	15.2	21	-0.056469	0.0201797	0.0954128	39.9824	40.629	40.34285768	-0.053461	15.22018
10	5.54	14.68	21.1	-0.011794	0.0488332	0.4197367				0.3147237	14.728833
11	5.25	14.86	20	0.0224372	0.0126656	0.5635268				0.6903139	14.872666
12	3.7	13.28	16	-0.060081	-0.0683	-0.197601				1.0492152	13.2117
13	4.4	13.26	18.5	0.0032881	0.004398	0.4116079				1.3250191	13.264398
14	4.4	13.58	18.4	0.0181458	-0.079471	-0.130074				1.6550889	13.500529
15	4.47	14.18	17	-0.050713	-0.092219	-0.233687				1.9739593	14.087781
16	4.8	15.3	17.6	0.0081208	-0.074748	-0.24097				2.2665866	15.225252
17	4.7	16.05	17.2	0.0022427	0.0575488	-0.26896				2.5695589	16.107549
18	4.66	15.68	17.5	-0.067172	0.0565313	-0.713103				2.8650613	15.736531
19		15.7	33.2		0.0276183					3.1580479	15.727618
Total	85.26	282.17	361.1	-0.202308	-0.185581	0.0262337					
Mean	4.7366667	14.851053	19.005263	-0.011239	-0.009767	0.0014574					
Pop.St.Dev	0.4372769	0.8605072	3.6572169	0.0364878	0.0559584	0.3883559					
							Ekregma				
							Sum of Angles plus Ekregma				

Table 3.1
An Extract of the Figure of Plan
Showing Some Principal Computed Statistics

BOSCAWEN-ÛN CIRCUIT (SEGMENTAL)

Area	635.1208
Perimeter	85.05769

CIRCLE

Radius	13.95282

ELLIPSE

z^2,$_x$	17.77153
M	407.1012
N	195.799
a	16.09933
b	12.16194
c	10.54874

CARTESIAN VALUES

Stone Number	Corrected Vertex Angle (radians)	Vertex Radius ρ_i	Boscawen-Ûn Stone Circuit Original (RRSZt) x	y	Restoration x	y	Circle x	y	Ellipse r(θ)	x	y
1	-2.563446	15.74321	-13.2009	8.57819	-13.2009	8.57819	-11.7332	7.62443	14.555564	-12.2051	7.3311
2	2.293768	15.15606	-10.0235	11.3602	-10.0235	11.3602	-9.25787	10.4924	13.456603	8.929519	10.1199
3	-1.97006	14.49086	-5.63318	13.3511	-5.63318	13.3511	-5.43957	12.8923	12.7626	-4.885909	-11.5076
4	-1.639747	13.87596	-0.68925	13.8588	-0.68925	13.8588	-0.69485	13.5756	12.16838	-0.60426	-12.1534
5	-1.300628	14.18908	3.78047	13.6744	3.78047	13.6744	3.734666	13.4852	12.35229	1.296809	-11.9042
6	0.862689	15.38876	8.797133	12.6263	8.797133	12.6263	7.999131	11.481	13.11685	7.698324	10.7623
7	-0.6462007	15.64868	12.48884	9.42921	12.48884	9.42921	11.16734	8.43147	14.2682	11.38711	-8.5974
8	-0.3561094	15.81478	14.82256	5.51351	14.82256	5.51351	13.11492	4.87832	15.4102	14.44337	-5.37246
9	0.0534607	15.22018	15.19843	0.81329	15.19843	0.81329	13.97283	0.74771	16.08207	16.05909	-0.85935
10	0.31477368	14.72883	14.00538	4.559365	14.00538	4.559365	13.39562	4.33153	15.54864	14.78492	4.81314
11	0.64933394	14.87267	11.46751	9.470605	11.46751	9.470605	10.78911	8.910938	14.09269	10.86612	8.973935
12	1.04931519	13.2117	6.587252	11.45497	6.587252	11.45497	6.971946	12.13223	12.86694	6.430934	11.15605
13	1.32501909	13.2644	3.277364	12.38578	3.277364	12.38578	3.404597	13.57232	12.1195	2.997461	11.94828
14	1.65508892	13.50853	-1.13865	13.45259	-1.13865	13.45259	-1.17809	13.94114	12.18049	1.02551	13.13724
15	1.97395928	14.08778	-5.32705	13.95829	-5.32705	13.95829	-5.48998	12.87994	12.58486	-4.93741	11.5787
16	2.26858665	15.22525	-9.75927	11.6861	-9.75927	11.6861	-8.596929	10.74015	13.4012	-8.59006	10.28606
17	2.56955888	16.10755	-13.5433	8.719712	-13.5433	8.719712	-11.7652	7.504918	14.57287	-12.2529	7.888924
18	2.86506127	15.78653	-15.1387	4.296395	-15.1387	4.296395	-13.4612	3.820326	15.66607	-15.0709	4.277157
19	3.15860479	15.72762	-15.7255	0.25879	-15.7255	0.25879	-13.9909	0.23024	16.09769	-16.0955	-0.26488

RADIAL ERROR

Vertex Radius ρ_i	Squared Vertex Radius ρ^2_i	Circle Point Radius r(circ)	Ellipse Point Radius r(ell)	Circle Squared Error r(circ)	Ellipse Squared Error r(ell)
15.74321	247.8487	13.99282	14.55564	3.063872	1.410316
15.15606	229.7044	13.99282	13.49603	1.359213	2.735828
14.49086	209.985	13.99282	12.7626	0.248042	2.665777
13.87596	192.5471	13.99282	12.16838	0.013657	2.915821
14.18908	201.33	13.99282	12.35229	0.038538	3.373813
15.38876	236.8138	13.99282	13.11685	1.948634	5.161569
15.64868	244.8811	13.99282	14.2682	2.741856	1.905717
15.81478	250.1072	13.99282	15.4102	3.319533	0.163687
15.22018	231.6539	13.99282	16.08207	1.50641	0.742852
14.72883	216.9385	13.99282	15.54864	0.541714	0.677087
14.87267	221.1962	13.99282	14.09269	0.774127	0.608556
13.2117	174.549	13.99282	12.86694	0.610149	0.11886
13.2644	175.9441	13.99282	12.1195	0.5306	0.898825
13.50853	182.2043	13.99282	12.18049	0.242351	1.742519
14.08778	198.4656	13.99282	12.58486	0.009038	2.258766
15.22525	231.8083	13.99282	13.4012	1.518886	3.327166
16.10755	259.4513	13.99282	14.57287	4.472075	2.355239
15.78653	247.6384	13.99282	15.66607	3.045527	0.004965
15.72762	247.358	13.99282	16.09769	3.009523	0.136955
19	19	19	19	19	19 Count
281.9944	4300.102	265.8636	263.3574	28.95671	34.19114 Sum
14.84129	221.0685	13.99282	13.86891	1.524669	1.799534 Mean
0.897091	26.26279	1.73E-15	1.362283	1.356119	1.423731 Pop.St.D
				0.006897	0.00814 Dimensionless MSE

Table 3.2
Cartesian Co-ordinates for Boscawen-Ûn Circuit Stones
and their
Circular and Elliptical Models

Simple Modus Paradigms Compared

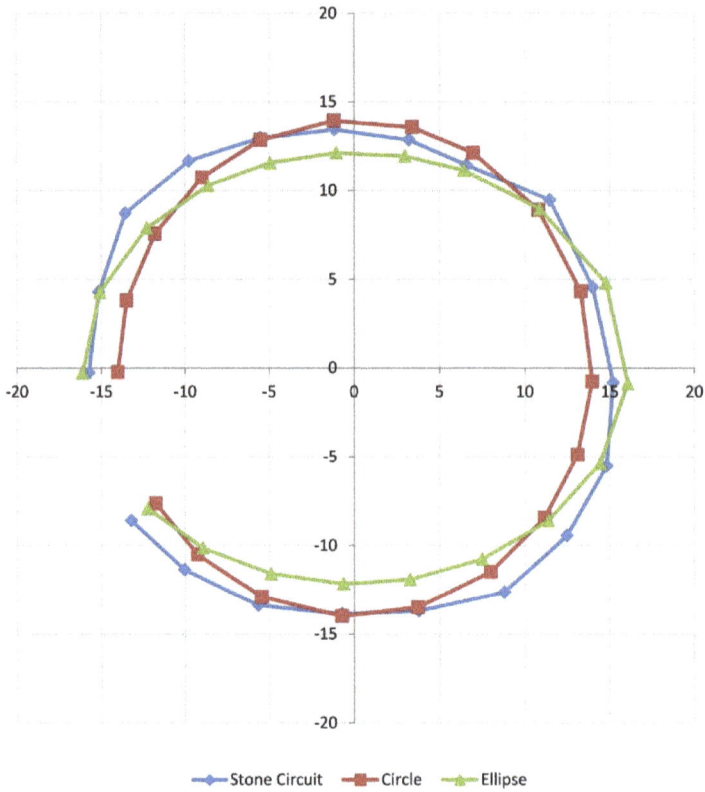

Figure 3.1
A Comparative Plot of the Boscawen-Ûn Stone Plan Data
and its
Model Circle and Ellipse

The Polar Profiles Compared

Table 3.3 lists the polar profile data for the Boscawen-Ûn stone circuit and its modelling circle and Ellipse, whilst Figure 3.2 displays the profiles graphically.

Corrected Vertex Angle (radians)	Vertex Radius ρ_i	Circle Point Radius r(circ)$_i$	Ellipse Point Radius r(elli)$_i$
-2.565345	15.74321	13.99282	14.55564
-2.293768	15.15006	13.99282	13.49603
-1.97006	14.49086	13.99282	12.57676
-1.620475	13.87596	13.99282	12.16838
-1.300623	14.18908	13.99282	12.35229
-0.962269	15.38876	13.99282	13.11685
-0.646701	15.64868	13.99282	14.2682
-0.356109	15.81478	13.99282	15.4102
-0.053461	15.22018	13.99282	16.08207
0.3147237	14.72883	13.99282	15.54864
0.6903139	14.87267	13.99282	14.09269
1.0492152	13.2117	13.99282	12.86694
1.3250191	13.2644	13.99282	12.3195
1.6550889	13.50053	13.99282	12.18048
1.9739593	14.08778	13.99282	12.58486
2.2665866	15.22525	13.99282	13.4012
2.5695589	16.10755	13.99282	14.57287
2.8650613	15.73653	13.99282	15.66607
3.1580479	15.72762	13.99282	16.09769

Table 3.3
Boscawen-Ûn Circuit Stones and their
Circular and Elliptical Models: Polar Profile Data

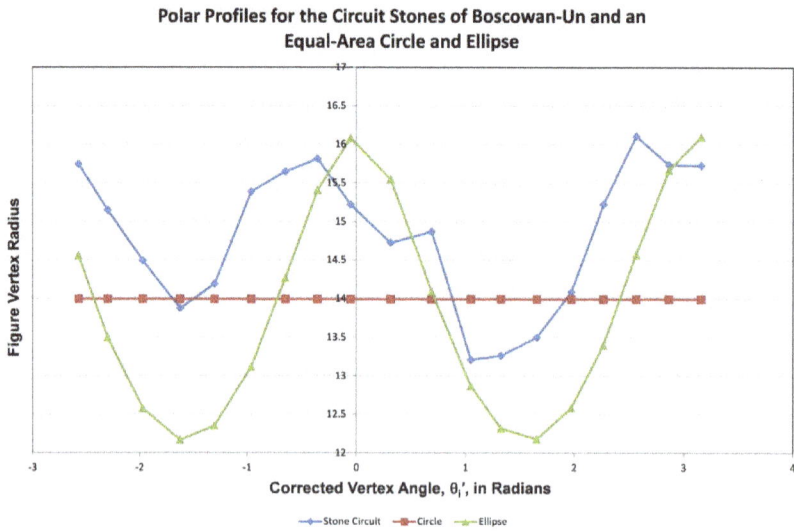

Polar Profiles for the Circuit Stones of Boscowan-Un and an Equal-Area Circle and Ellipse

**Figure 3.2
Boscawen-Ûn Circuit Stones and their
Circular and Elliptical Models: Polar Profile Data Plot**

As one might expect the Circle polar profile flatlines at a constant radius, whilst the Ellipse profile is distinctly cyclical. The ellipse profile shows marked overshoot relative to the Boscawen-Ûn line which substantially accounts for its high squared error.

The Mean Squared Error

In the light of the above, reference to Table 3.2 shows that Circle Mean Squared Error (MSE) is 0.006896815 and Ellipse MSE is 0.008140164 (value rounded in the table). This may seem counter-intuitive. But we must remember that aerial photographs confirm that the ground plan of the Boscawen-Ûn stone circuit is *very manifestly neither a circle nor an ellipse*, and it would presumably be impossible to draft a planar envelope of a single dipole or monopole figure using a rope loop.

Curvature Discontinuity

We read in Chapter Two that the Boscawen-Ûn ground plan seemed to separate into four sub-arcuate sectors each of five stones, except that the western sector had only four stones, which may infer the removal of one of the stones in the remote past to form a wide gap.

In Chapter Four we shall evaluate a simple four-part scheme for forming a figure of four orthogonal arcs.

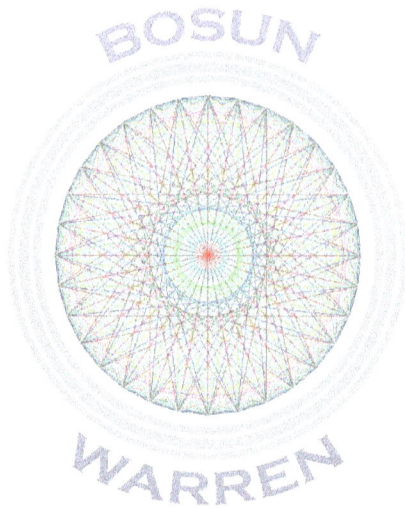

THE INADEQUACY OF MONOPOLAR
MODUS MODELS
(ATTEMPTS TO FIT CIRCULAR ARCS)

Inspection of the plan of the stone circuit of Boscawen-Ûn makes manifest that the layout is in no manner a circle.

Notwithstanding, we shall fit circles to groups of three peripheral stones in order decisively to demonstrate the fallacy.

A circle of unique radius and position is generated by any three points on its periphery.

To fix the Center of the Circle, x_c and y_c; and also the Radius r we need these equations:-

$$x_c = -\left(\frac{\theta}{2\mu}\right)$$
Equation 4.1

$$y_c = -\left(\frac{\rho}{2\mu}\right)$$
Equation 4.2

$$r = \frac{\sqrt{\theta^2 + \rho^2 - 4\mu\sigma}}{4\mu^2}$$
Equation 4.3

where:-

$$\mu = x_1(y_2 - y_3) - y_1(y_2 - x_3) + x_2 y_3 - x_3 y_2$$
Equation 4.4

$$\theta = (x_1^2 + y_1^2)(y_3 - y_2) + (x_2^2 + y_2^2)(y_1 - y_3) + (x_3^2 + y_3^2)(y_2 - y_1)$$
Equation 4.5

$$\rho = (x_1^2 + y_1^2)(x_2 - x_3) + (x_2^2 + y_2^2)(x_3 - x_1)$$
$$+ (x_3^2 + y_3^2)(x_1 - x_2)$$
Equation 4.6

$$\sigma = (x_1^2 + y_1^2)(x_3 y_2 - x_2 y_3) + (x_2^2 + y_2^2)(x_1 y_3 - x_3 y_1)$$
$$+ (x_3^2 + y_3^2)(x_2 y_1 - x_1 y_2)$$
Equation 4.7

Three of the four sectors contain five stones and a fourth sector only four.

The number of possible Combinations, ^nC_r, of three stones, r, chosen from five stones, n, is given by:-

$$C_r^n = \frac{n!}{r!\,(n-r)!} = \frac{5!}{3!\,(5-3)!} = \frac{120}{6 \times 2} = 10$$
Equation 4.8

If we code the five stone vertices of Sector One as A, B, C, D and E we may elaborate the ten combinations and their ten resulting sets of $\{x_c, y_c, r\}$ as presented in Table 4.1:-

It is immediately apparent, even without plotting, that the large standard deviation of the y_c values betrays a sub-linear disposition of the fitted circle centers: Something that would be impossible if circles fitted the stones. If circles fitted the stones then the ten centers would present ideally as a single point, but more realistically as a nucleated cloud of points in 2D space, due to statistical error.

The Curvature, κ_c, of a Circle is given by:-

$$\kappa_c = \frac{1}{r}$$
Equation 4.9

Since Sector One radii vary from 15.25845931 to 24.59683962 the curvature of Sector One's trace varies as their reciprocals: Clearly radius should be quasi-constant. Similar arguments may be educed for the other three sectors.

For the sake of completeness I offer the Center-and-Radius outcomes for all four sectors as Tables 4.2a and 4.3b.

The sub-linearity of the circle-center loci for all the sectors is also illustrated in the planimetric plot of Figure 4.1. Figure 4.2 is the same with Circles added that fit the central and extreme three stones of each sector.

Sector	Stone	By-Eye RRSZI x	y	Vertex Code
1	12	6.582752	11.45497	A
1	13	3.227364	12.86578	B
1	14	-1.13665	13.45259	C
1	15	-5.52705	12.95829	D
1	16	-9.75927	11.6861	E

Combinations ^tC_r	Chosen Vertex U x	y	Chosen Vertex V x	y	Chosen Vertex W x	y	Algebraic Intermediate μ	θ	ρ	σ	Circle Center CoOrdinates x_c	y_c	Radius of Circle r_c
ABC	6.582752	11.45497	3.227364	12.86578	-1.13665	13.45259	4.187811	8.097612	15.11732	-957.451	-0.966807309	-1.804919248	15.25845931
ABD	6.582752	11.45497	3.227364	12.86578	-5.52705	12.95829	12.04043	31.64425	63.35336	-3035.66	-1.314083311	-2.630859971	16.14839591
ABE	6.582752	11.45497	3.227364	12.86578	-9.75927	11.6861	22.27996	80.45946	169.3261	-6358.22	-1.805646453	-3.799964703	17.40914543
ACD	6.582752	11.45497	-1.13665	13.45259	-5.52705	12.95829	12.58612	36.17778	91.19128	-3479.64	-1.437209886	-3.62269333	17.0779332
ACE	6.582752	11.45497	-1.13665	13.45259	-9.75927	11.6861	30.861	112.5991	315.9243	-9746.87	-1.824294813	-5.118503476	18.58382496
BCD	3.227364	12.86578	-1.13665	13.45259	-5.52705	12.95829	4.733499	12.63114	42.95524	-1426.25	-1.334228931	-4.537366555	17.99104633
BCE	3.227364	12.86578	-1.13665	13.45259	-9.75927	11.6861	12.76885	40.23728	161.7155	-4457.06	-1.575602883	-6.33242043	19.78988402
DAE	-5.52705	12.95829	6.582752	11.45497	-9.75927	11.6861	-21.7683	-80.551	-302.554	7795.626	-1.850189682	-6.949409262	20.24440128
DBE	-5.52705	12.95829	3.227364	12.86578	-9.75927	11.6861	-11.5288	-31.7358	-196.581	4660.01	-1.376372396	-8.525667507	21.88123733
DCE	-5.52705	12.95829	-1.13665	13.45259	-9.75927	11.6861	-3.49341	-4.12961	-77.8206	1678.92	-0.593057046	-11.13819312	24.59683962
Mean											-1.407549271	-5.44599976	18.89811674
Pop.SD											0.378577745	2.70334285	2.669118066

Table 4.1
The Centers and Radii of the Ten Possible Circles
Derivable from the Sector One Stone Co-Ordinates

SECTOR 1

Stone	By-Eye RRSZI x	y	Vertex Code
12	6.582752	11.45497	A
13	3.227364	12.86578	B
14	-1.13665	13.45259	C
15	-5.52705	12.95829	D
16	-9.75927	11.6861	E

Combinations ^nC_r	Chosen Vertex U		Chosen Vertex V		Chosen Vertex W		Algebraic Intermediate				Circle Center CoOrdinates		Radius of Circle
	x	y	x	y	x	y	μ	θ	ρ	σ	x_c	y_c	r_c
ABC	6.582752	11.45497	3.227364	12.86578	-1.13665	13.45259	4.187811	8.097612	15.11732	-957.451	-0.966807309	-1.804919248	15.25845931
ABD	6.582752	11.45497	3.227364	12.86578	-5.52705	12.95829	12.04043	31.64425	63.35336	-3035.66	-1.314083311	-2.630859971	16.14839591
ABE	6.582752	11.45497	3.227364	12.86578	-9.75927	11.6861	22.27996	80.45946	169.3261	6358.22	-1.805646453	-3.799964703	17.40914543
ACD	6.582752	11.45497	-1.13665	13.45259	-5.52705	12.95829	12.58612	36.17778	91.19128	3479.64	-1.437209866	-3.62269333	17.0779332
ACE	6.582752	11.45497	-1.13665	13.45259	-9.75927	11.6861	30.861	112.5991	315.9243	-9746.87	-1.824294813	-5.118503476	18.58382496
BCD	3.227364	12.86578	-1.13665	13.45259	-5.52705	12.95829	4.733499	12.63114	42.95524	-1426.25	-1.334228931	-4.537366555	17.99104633
BCE	3.227364	12.86578	-1.13665	13.45259	-9.75927	11.6861	12.76885	40.23728	161.7155	-447.06	-1.575602883	-6.33242043	19.78988402
DAE	-5.52705	12.95829	6.582752	11.45497	-9.75927	11.6861	-21.7683	-80.551	-302.554	7795.626	-1.850189682	-6.949409262	20.24440128
DBE	-5.52705	12.95829	3.227364	12.86578	-9.75927	11.6861	-11.5288	-31.7358	-196.581	4660.01	-1.376377396	-8.525667507	21.88123733
DCE	-5.52705	12.95829	-1.13665	13.45259	-9.75927	11.6861	-3.49341	-4.12961	-77.8206	1678.92	-0.591057046	-11.13819312	24.59683962
Mean											-1.407549271	-5.44599976	18.89811674
Pop.SD											0.378577245	2.70334285	2.669118066

SECTOR 2

Stone	By-Eye RRSZI x	y	Vertex Code
17	-13.5433	8.719712	A
18	-15.1387	4.296395	B
19	-15.7255	-0.25879	C
1	-13.2009	-8.57819	D

Combinations ^nC_r	Chosen Vertex U		Chosen Vertex V		Chosen Vertex W		Algebraic Intermediate				Circle Center CoOrdinates		Radius of Circle
	x	y	x	y	x	y	μ	θ	ρ	σ	x_c	y_c	r_c
ABC	-13.5433	8.719712	-15.1387	4.296395	-15.7255	-0.25879	4.671747	-52.5777	6.48565	-1980.73	5.627202063	-0.694135467	21.35712861
ABD	-13.5433	8.719712	-15.1387	4.296395	-13.2009	-8.57819	29.11178	-153.04	-22.5588	-9429.09	2.628486973	0.387451701	18.19208056
ACD	-13.5433	8.719712	-15.7255	-0.25879	-13.2009	-8.57819	40.82202	-105.031	-29.4645	-11756.9	1.286447151	0.360089427	17.02321871
BCD	-15.1387	4.296395	-15.7255	-0.25879	-13.2009	-8.57819	16.38199	-4.56862	-0.42001	-4124.17	0.139440348	0.012819128	15.8672542
Mean											2.420394134	0.016756197	18.10992052
Pop.SD											2.050335683	0.43624113	2.047045659

Table 4.2a
The Centers and Radii of the Possible Circles Derivable from the Sector One and Sector Two Stone Co-Ordinates

SECTOR 3

By-Eye RRSZI			Vertex Code
Stone	x	y	
2	-10.0235	-11.3602	A
3	-5.63318	-13.3511	B
4	-0.68905	-13.8588	C
5	3.787047	-13.6744	D
6	8.797133	-12.6263	E

Combinations °C_r	Chosen Vertex U x	y	Chosen Vertex V x	y	Chosen Vertex W x	y	Algebraic Intermediate μ	θ	ρ	σ	Circle Center CoOrdinates x_c	y_c	Radius of Circle r_c
ABC	-10.0235	-11.3602	-5.63318	-13.3511	-0.68905	-13.8588	7.614431	24.80725	-20.0253	-1726.53	-1.62896278	1.314958306	15.2028776
ABD	-10.0235	-11.3602	-5.63318	-13.3511	3.787047	-13.6744	17.336	10.91555	-146.068	-5528.98	-0.31482317	4.212838236	18.35149584
ABE	-10.0235	-11.3602	-5.63318	-13.3511	8.797133	-12.6263	31.91205	-39.2528	-399.748	-12259.3	0.615015829	6.26327622	20.58553673
ACD	-10.0235	-11.3602	-0.68905	-13.8588	3.787047	-13.6744	12.90622	-15.1358	-247.566	-5926.4	0.586374866	6.590971937	23.48446789
ACE	-10.0235	-11.3602	-0.68905	-13.8588	8.797133	-12.6263	35.2076	-65.0393	-764.073	-17412.9	0.923653709	10.85097143	24.76237932
BCD	-5.63318	-13.3511	-0.68905	-13.8588	3.787047	-13.6744	3.184652	-1.24407	-121.524	-2298.22	0.195322285	19.07966716	32.95037401
BCE	-5.63318	-13.3511	-0.68905	-13.8588	8.797133	-12.6263	10.90998	-0.97919	-384.351	-7427.96	0.044875792	17.61463398	31.48202662
DAE	3.787047	-13.6744	-10.0235	-11.3602	8.797133	-12.6263	-26.0683	52.56774	631.3081	13681.99	1.008270564	12.10874426	25.93241893
DBE	3.787047	-13.6744	-5.63318	-13.3511	8.797133	-12.6263	-11.4922	2.39386	377.6276	7468.46	0.104390642	16.42970692	30.32848725
DCE	3.787047	-13.6744	-0.68905	-13.8588	8.797133	-12.6263	-3.76689	2.664239	114.801	2318.129	0.353639329	15.23818331	29.11569631
Mean											0.188775707	11.77039518	25.21957605
Pop.SD											0.718625018	5.668211504	5.594382175

SECTOR 4

By-Eye RRSZI			Vertex Code
Stone	x	y	
7	12.48884	-9.42921	A
8	14.82256	-5.51351	B
9	15.19843	-0.81329	C
10	14.00538	4.559365	D
11	11.46751	9.470605	E

Combinations °C_r	Chosen Vertex U x	y	Chosen Vertex V x	y	Chosen Vertex W x	y	Algebraic Intermediate μ	θ	ρ	σ	Circle Center CoOrdinates x_c	y_c	Radius of Circle r_c
ABC	12.48884	-9.42921	14.82256	-5.51351	15.19843	-0.81329	9.497244	-96.8221	45.02956	-691.906	5.097380119	-2.370664614	10.2204065
ABD	12.48884	-9.42921	14.82256	-5.51351	14.00538	4.559365	26.70722	-182.522	73.13607	-3570.99	3.417082544	1.369219132	12.13508105
ABE	12.48884	-9.42921	14.82256	-5.51351	11.46751	9.470605	48.10626	-191.517	49.93646	-8917.63	1.990561091	-0.519022426	13.76972719
ACD	12.48884	-9.42921	15.19843	-0.81329	14.00538	4.559365	24.83701	-55.721	55.65343	-4861.45	1.121733794	-1.120373024	14.08005007
ACE	12.48884	-9.42921	15.19843	-0.81329	11.46751	9.470605	60.01058	45.92447	77.68577	-14536.5	-0.382636442	-0.647267195	15.58195584
BCD	14.82256	-5.51351	15.19843	-0.81329	14.00538	4.559365	7.627033	29.97839	27.54692	-2200.05	-1.965272155	-1.8058738	17.19238355
BCE	14.82256	-5.51351	15.19843	-0.81329	11.46751	9.470605	21.40157	140.6193	72.77887	-7035.76	-3.285256358	-1.700316437	18.50496282
DAE	14.00538	4.559365	12.48884	-9.42921	11.46751	9.470605	-42.9493	-196.791	-644.576	12367.38	-2.290967605	-0.750399138	17.13955988
DBE	14.00538	4.559365	14.82256	-5.51351	11.46751	9.470605	-21.5503	-205.786	-87.6572	7956.863	-4.774564951	-2.033783495	19.90366815
DCE	14.00538	4.559365	15.19843	-0.81329	11.46751	9.470605	-7.7574	-95.1455	-42.4252	3212.839	-6.118097936	-2.728050061	21.40235645
Mean											-0.71900379	-1.504496187	15.99301465
Pop.SD											3.434944437	0.713945183	3.325585868

Table 4.2b
The Centers and Radii of the Possible Circles
Derivable from the Sector Three and Sector Four Stone Co-
Ordinates

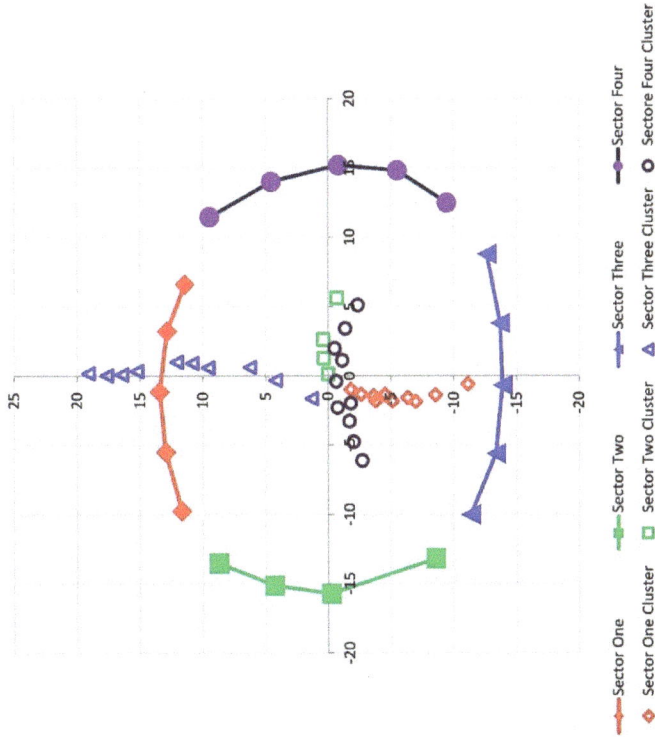

Figure 4.1
Plan of the Boscawen-Ûn Stone Sectors
Showing the Quasi-Linear Disposition of Fitted Circle Centers

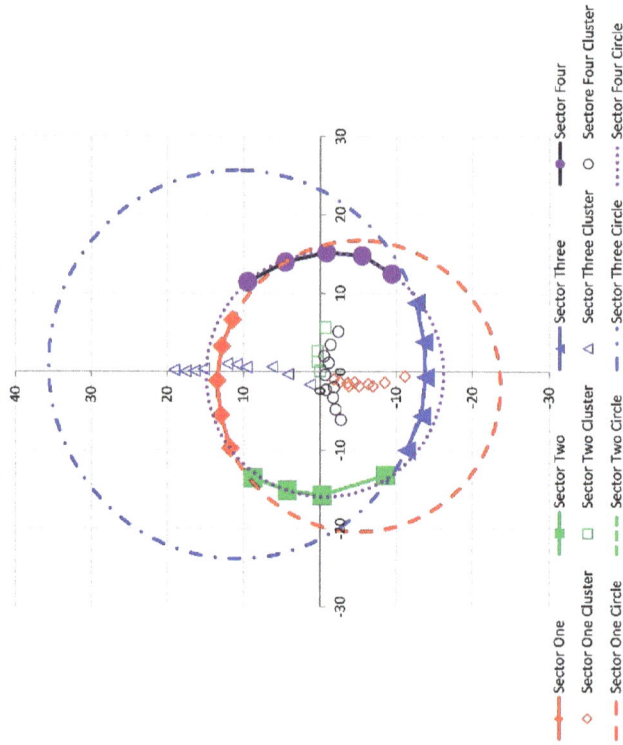

Figure 4.2
Plan of the Boscawen-Ûn Stone Sectors
Showing the Quasi-Linear Disposition of Fitted Circle Centers
And Also Circles Representing the Central and Extreme Points
of Each Sector

CHAPTER FIVE
ASPECTS OF THE ANALYTIC GEOMETRY
OF OFFSET ELLIPSES

Figure 5.1 shows an ideal ellipse drafted with its major radii at the upper left and upper right vertices of a rectangle.

We are investigating this conformation because our chief thesis is the Ancients drafted the trace of the Boscawen-Ûn circuit stones using a rope loop stretched around two focal turnposts at the top of a rectangle of four vertex turnposts. Such a procedure generates an elliptical arc above and between the upper turnposts.

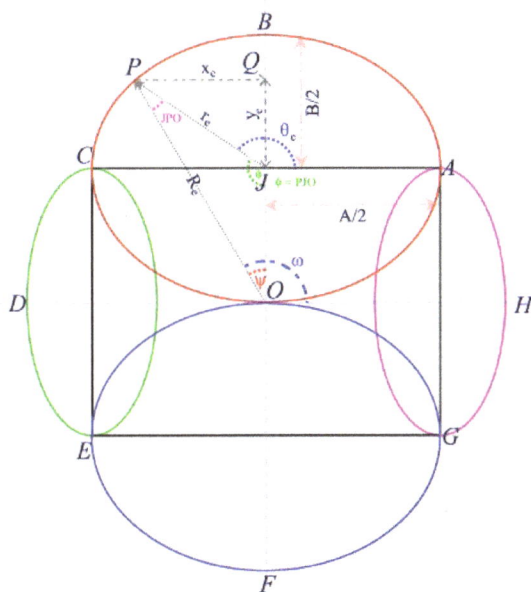

Figure 5.1
Four Offset Ellipses
whose Interfocal Distances 2c are
The Sides of a Rectangle

The green, Western ellipse is the ellipse that fits selected Western stone positions; the blue, Southern ellipse that for

Southern stones and the purple, Eastern ellipse the fit for Selected Eastern stones. Every stone is included in one of the ellipses.

The red Ellipse *ABCO* is the Northern, Sector One stone-fitting ellipse.

The *ABCO* Interradial Distance *CA* is the upper side of Rectangle *ACEG* (i.e. it is *CA*).

Some Point *P* on the upper arc of *ABCO* is Fit Radius R_e from the General Center *O*: And *P* is Ellipse Radius r_e from the local Ellipse Center *J*.

(Note that in Chapter Three we have $R_e = \rho$).

Line *JA* = A/2 where in context A is the Ellipse Major Semi-Axis, a, and in the same spirit *JB* = B/2 is the Ellipse Minor Semi-Axis, b.

Now the Eccentricity, e, of the ellipse is given by:-

$$e = \frac{c}{a} = \sqrt{1 - \frac{b^2}{a^2}}$$

Equation 5.1

where c is the Distance between a Focus and the Ellipse Center; a is the Major Semi-Axis; and b is the Minor Semi-Axis.

whilst the Ellipse Radius r_e is given by:-

$$r_e = \frac{ab}{\sqrt{(b.\cos\theta)^2 + (a.\sin\theta)^2}} = \frac{b}{\sqrt{1 - (e.\cos\theta)^2}}$$

Equation 5.2

where r_e is the Radius Angle.

If x_e is the Cartesian Abscissal Co-Ordinate of *P*, and y_e is Ordinal Co-Ordinate, then:-

$$x_e = r_e.\cos\theta$$
Equation 5.3a

$$y_e = r_e.\sin\theta$$
Equation 5.3b

Note that since we are treating of the upper half of an ellipse with $0 \leq \theta \leq \pi$ we have $\cos(\theta)$ = $-\cos(\pi-\theta)$ and $\sin(\theta)$ = $\sin(\pi-\theta)$ and accordingly Equations 5.3 do not modify for quarter.

Accordingly we may write:-

$$x_e = \frac{b}{\sqrt{1-(e.\cos\theta)^2}}\cos\theta$$

Equation 5.4a

$$y_e = \frac{b}{\sqrt{1-(e.\cos\theta)^2}}\sin\theta$$

Equation 5.4b

The statement of the trace-drawing modus implies that:-

$$Rectangle\ Width = A$$

Equation 5.5a

$$Rectangle\ Height = B$$

Equation 5.5b

where A and B are respectively the Major Axis length and B the Minor Axis of the ellipse.

R_e is the Vertex Radius, the distance between the circuit barycenter at O and the Position P, which is ideally a stone position.

Therefore, for the unshifted conformation of Figure 5.1:-

$$R_e = \sqrt{r_e{}^2 + \frac{B^2}{2} - 2.r_e.\frac{B}{2}.\cos\left(\frac{6}{4}\pi - \theta\right)}$$

Equation 5.6

Setting the angle PJO to ϕ we may write:-

$$\phi = PJO = \frac{6}{4}\pi - \theta = \frac{3}{2}\pi - \theta$$
Equation 5.7

In order to generalise ϕ for the North-East as well as the North-West quadrant of the ellipse it is convenient to establish the auxiliary parameter U as:-

$$U = entier\left(\frac{\theta}{\frac{\pi}{2}}\right) = entier\left(\frac{2.\theta}{\pi}\right)$$
Equation 5.8

whilst:-

$$\phi = PJO = \frac{2U + 1}{2}\pi + (-1)^U.\theta$$
Equation 5.9

Re-arrangement confirms Equation 5.6 to be the square root of a quadratic equation in r_e:-

$$R_e = \sqrt{r_e^2 - 2.r_e.\frac{B}{2}.\cos(\phi) + \frac{B^2}{2}}$$
Equation 5.10

The (complex) roots $R1$ and $R2$ of this quadratic are:-

$R1$

$$= \frac{-\left(-2.\left(\frac{B}{2}\right).\cos\phi\right) + \sqrt{\left(-2.\left(\frac{B}{2}\right).\cos\phi\right)^2 - 4.1.\left(\frac{B}{2}B/2\right)^2}}{2.1}$$
Equation 5.11a

$$R2$$

$$= \frac{-\left(-2.\left(\frac{B}{2}\right).\cos\phi\right) - \sqrt{\left(-2.\left(\frac{B}{2}\right).\cos\phi\right)^2 - 4.1.\left(\frac{B}{2}B/2\right)^2}}{2.1}$$

Equation 5.11b

Elementary simplifications of these quadratic root equations give:-

$$R1 = \frac{B}{2}\left[\cos\phi + \sqrt{\cos\phi^2 - 1}\right] = \frac{B}{2}\left[\cos\phi + \sqrt{-(-\sin\phi)^2}\right]$$

Equation 5.12a

$$R2 = \frac{B}{2}\left[\cos\phi - \sqrt{\cos\phi^2 - 1}\right] = \frac{B}{2}\left[\cos\phi - \sqrt{-(\sin\phi)^2}\right]$$

Equation 5.12b

Therefore the Vertex Radius R_e may be re-quoted as:-

$$R_e = \sqrt{1.(r_e - R1).(r_e - R2)} = \sqrt{(r_e - R1).(r_e - R2)}$$

Equation 5.13

which is real.

With regard to the North-East Quadrant, the conformation of Figure 5.2 pertains:-

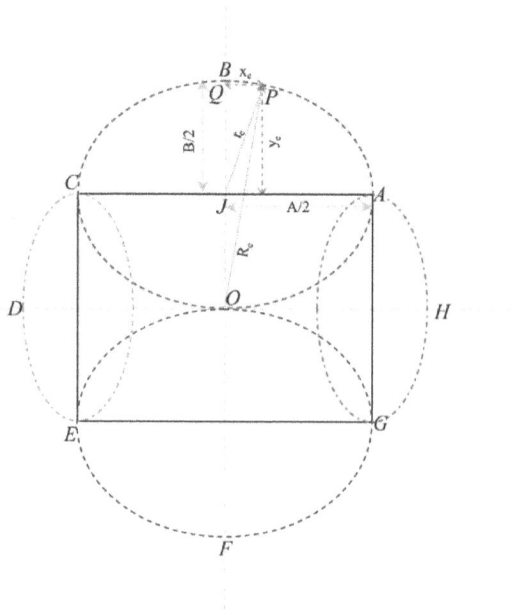

Figure 5.2
Four Offset Ellipses
whose Interfocal Distances 2c are
The Sides of a Rectangle

Once again PJA is ω and PJO is ϕ. The algebra is the same as for the North-West quadrant example above.

A Development for an Offset Fitment Ellipse

Figure 5.3 shows the conformation of an ellipse *ABCO* (blue dashed) whose Interfocal Base f1 to f2 is identical to the Northern Interpost and Interfocal Distance shorter than *CA* = A.

That Interfocal distance 2c is represented by the broad green line centered at *J*.

The red dashed ellipse is identical to ABCO, but shifted to the South-East by x-Shift xs (nominally +7) and y-Shift ys which is by example -8.8.

p1 through p5 are points randomly dispersed near to the two ellipses, and to their diagrammatic North. p1 through p5 represent circuit stones of Boscawen-Ûn.

Each Point p stands at the end of a (gray) ray that converges to General Barycenter O. Each ray respectively intercepts the shifted red ellipse at q1 through q5. Of course, all these point rays converge at O.

The blue co-ordinates of p1...p5 and q1...q5 have been estimated from the PhotoDraw source diagram for purposes of checking my algebra.

You have possibly ascertained that knowledge of the five straight-line distances between p_i ... q_i for i = 1...5 will enable the computation of basic statistical measures of ellipse fit quality to the stone points.

One such basic metric is the Root Mean Square Error, $RMSE_{pq}$, defined by:-

$$RMSE_{pq} = \sqrt{\left|\frac{\sum_{i=1}^{5}(p_i - q_i)^2}{5}\right|}$$

Equation 5.14

Or generally:-

$$RMSE_{pq} = \sqrt{\left|\frac{\sum_{i=1}^{n}(p_i - q_i)^2}{n}\right|}$$

Equation 5.15

where n is the number of p-q Point Pairs.

The RMSE is ideally zero but in practical terms the smaller is the RMS error the better the fit of the ellipse to the stone points.

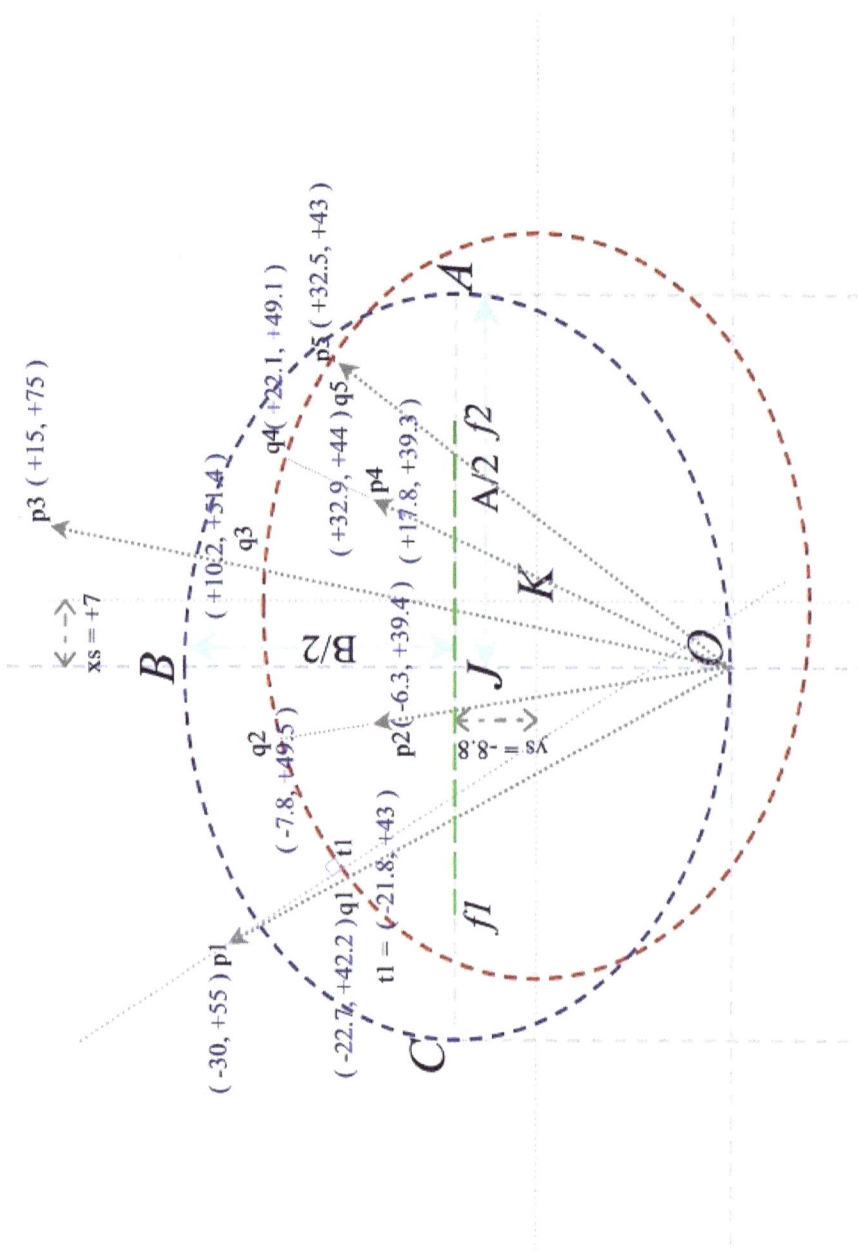

Figure 5.3
Unshifted and Shifted Fitment Ellipse Geometry

The Minimum Possible Closed Loop Length for generating a figure about the rectangle A×B is given by:-

$$L = A + 2(a + B)$$
Equation 5.16

This expression yields 280 when A = 80 and B = 60.

In practice, ancient men would have used a slightly longer loop to geoscribe Boscawen-Ûn, but not the kind of loop-length that would have given the ellipses of Figure 5.3, which is a caricature composed for illustrative purposes.

The Algebraic Establishment of the Interception Points, q_i [5.1, 5.2]

For computational purposes let us identify px_i as the x co-ordinate of Point p_i and py_i as its y co-ordinate: Similarly, qx_i is the x co-ordinate of Crossing q_i and qy_i its y co-ordinate.

Before we actually fix the Crossings co-ordinates it is convenient to define some auxiliary parameters.

Because we shall only deal with the upper quadrants of the ellipses the ordinate is always positive and the abscissa negative in the North-West and the positive in the North-East. Hence:-

$$Sgn_i = sign(px_i)$$
Equation 5.17

Also, the X-Shift Parameter, h_e, is given by:-

$$h_e = x_0 + xs$$
Equation 5.18a

and the Y-Shift Parameter, k_e, is given by:-

$$k_e = y_0 + ys$$
Equation 5.18b

In our system of Figure 5.3 (x_0, y_0) is the co-ordinate if the unshifted ellipse center J which is $(0,30)$. Therefore, $h_e = +7$ and $k_e = +21.2$.

A further auxiliary is:-

$$\varphi_i = \kappa_i - k_e$$
Equation 5.19

where κ_i is the Line Intercept of the straight-line containing p and q. Now in our system κ_i is invariably zero and effectively κ_i is $-k_e$ or -21.2

Our fifth convenience auxiliary is the Line Gradient, g_i:-

$$g_i = \frac{py_i}{px_i}$$
Equation 5.20

There is no need to deduct the lower co-ordinate values because the lines converge to $(0,0)$ by definition.

Our sixth and final auxiliary is x-Axis Crossing, μ_i:-

$$\mu_i = \kappa_i + g_i . h_e$$
Equation 5.21

We are now in a position to define the Intersection sets of qx_i and qy_i:-

$$
\begin{aligned}
&qx_i \\
&= \frac{b^2 . h_e - a^2 . g_i^2 . \varphi_i + Sgn_i . a . b . \sqrt{b^2 + a^2 . g_i^2 - 2 . g_i . \varphi_i . h_e - \varphi_i^2 - g_i^2 . h_e^2}}{b^2 + a^2 . g_i^2}
\end{aligned}
$$
Equation 5.22

$$
qy_i = \frac{b^2 . \mu_i + a^2 . g_i^2 . k_e + Sgn_i . a . b . g_i . \sqrt{b^2 + a^2 . g_i^2 + 2 . \mu_i . k_e - k_e^2 - \mu_i^2}}{b^2 + a^2 . g_i^2}
$$
Equation 5.23

To establish RMS error it is now convenient to use the Pythagorean Theorem to compute the point ray Line Segments R_i, the Total Radial Length from the origin at $O \equiv (0,0) \equiv (x_0,y_0)$ to p_i: And S_i, the Radial Length from p_i to q_i:-

$$R_i = \sqrt{(px_i - x_0)^2 + (py_i - y_0)^2} = \sqrt{(px_i)^2 + (py_i)^2}$$
Equation 5.24

$$S_i = \sqrt{(px_i - qx_i)^2 + (py_i - qy_i)^2}$$
Equation 5.25

And Deviation, δ_i, is given by:-

$$\delta_i = R_i - S_i$$
Equation 5.26

An available expression of RMS error is now resolvable as:-

$$RMSE_{pq} = \sqrt{\left|\frac{\sum_{i=1}^{n}(R_i - S_i)^2}{n}\right|} = \sqrt{\left|\frac{\sum_{i=1}^{n}\delta_i^2}{n}\right|}$$
Equation 5.27

Precision

Unfortunately, the quality of both the data and my analysis preclude high levels of precision.

We can firstly check the integrity of our methods by certifying that EXCEL® and MathCad® agree with one another. Both are nominally accurate to fifteen figures and accordingly we anticipate that the RMSE (Root Mean Squared Error) of their PSDs (Percentage Specific Defects) would approximate 10^{-7}.

On the other hand, when we engage material derived from my idealised test plot FigureThreeVcrop.jpg things are much worse due to human error, and RMSEs in the region of 10^0 become commonplace, reflecting PSDs that range typically from a fraction of a percentage, to five percent.

Once again, we can only hope that our methods *substantiate the principle* of our theses about the objective world.

Science is a civil action: Its findings hinge on a balance of probabilities, rather than proof beyond reasonable doubt.

Table 5.1 presents the algebraic parameters that define the geometry of our Ideal Test Ellipse as shown in its unshifted and shifted forms in Figure 5.3

It is these parameters that are used in the spreadsheet and scratchpad parallel elaborations.

The first of these EXCEL® elaborations is given in Table 5.2

Consult Cell N41 and O41, respectively the RMSEs of the PSDs of the Cartesian x and y of the Points q_1...q_5 and also Point t_1.

The value for x is 2.26924E-07 (i.e. 2.26924×10^{-7}) proving beyond reasonable doubt that:-

(A)   The Intersection of The Ellipse Methodology advanced by the AmbrSoft® Process is mathematically sound.

(B)   The accuracy of elaboration of both 64-bit MathCad® and 64-bit EXCEL® is fifteen digits of decimal as (implicitly) claimed by their manufacturers.

The value for y is 0.63256492 ($\sim 10^0$) demonstrating that large error has insinuated into the process in the computation of ordinal displacements. This is due to the error of hand and eye inserted by the human agent during the process of translating aerial photograph data to three-figure accuracy via the graphics of Google®, PhotoDraw® and desktop hardware.

Therefore, the total analysis can be no better than an average of PSDs lying randomly between zero and ten percent (hopefully, five percent).

I do not think I need to spell-out the implications for modern science, even for my younger readers.

Ideal Ellipse Parameters

A	80
B	60
a	40
b	30
ci	0
xc	0
yc	0
xs	7
ys	-8.8
x0	0
y0	30
ci	0
he	7
ke	21.2
φ	-21.2

Table 5.1
Ideal Ellipse Parameters

Table 5.3 presents the precision of the calculation with especial reference to the Radial Traces of the Ideal Test Ellipse.

We treat of lines between the Origin and Points $q_1...q_5$ and also Point t_1, except for the Plot Radii themselves.

Consult the RMSEs arrayed in L41 to R41

By now propagative errors have polluted the entire system to the extent of generating RMSEs lying between 0.462484951 for the PSDs of Plot Lines to 6.543028357 for Radial Line Lengths computed by EXCEL.

Note especially that the computation of radii from computations increases error by about 25%.

Table 5.4 shows Deviations in Columns N to P. These deviations are the origin-referent distances δ between the Points $p_1...p_5$ and their respective Ellipse Interceptions $q_1...q_5$.

The RMSEs of the (N41) Plot Deviations δ and the (O41) EXCEL Deviations are uncannily similar at respective values of 3.487354339 and 3.479751468.

The RMSE of the PSDs of the Deviations is 1.06293675 and is displayed in Cell P41.

Serial	Point	RADIAL LINE LENGTH — PLOT From Origin Zero	RADIAL LINE LENGTH — COMPUTED From Origin Zero	POINTS Plot x	POINTS Plot y	Computed (MathCad) x	Computed (MathCad) y	Computed (EXCEL) x	Computed (EXCEL) y	PSDs OF POINTS MathCad vs. EXCEL x	PSDs OF POINTS MathCad vs. EXCEL y	PSDs OF POINTS Plot vs. EXCEL x	PSDs OF POINTS Plot vs. EXCEL y	PSD Of Plot Lines	RADIAL LINE SQUARED ERROR FRACTION
1	p1	62.62	62.64982043	-30	55									-0.047621256	0.052293754
2	p2	40	39.90050125	-6.3	39.4									0.248746875	0.065481446
3	p3	76.85	76.4852927	15	75									0.47450327	0.099143764
4	p4	43.13	43.14313387	17.8	39.3									-0.030451817	0.061524401
5	p5	54.13	53.90037106	32.5	43									0.42421752	0.000377068
6	q1	47.91795071		-22.7	42.2	-22.57925864	41.39530751	-22.57925864	41.39530751	-6.29376E-14	3.43296E-14	0.531900264	1.906854249		
7	q2	50.11077728		-7.8	49.5	-7.844253334	49.05771133	-7.844253334	49.6783852	-3.3968E-14	-1.265191285	-0.567350037	-0.360374174		
8	q3	52.40229003		10.2	51.4	10.22052134	51.10260668	10.22052134	51.10260668	-3.47606E-13	4.17127E-14	-0.20118957	0.578586224		
9	q4	53.84440547		22.1	49.1	22.17415714	48.95754919	22.17415714	47.35865283	1.44197E-13	3.265883176	-0.335552674	3.546531913		
10	q5	54.94005825		32.9	44	33.17095788	43.88772889	33.17095788	43.88772889	4.28413E-14	-0.823580182	-0.823580182	0.25516162		
11	t1	48.21037233		-21.8	43			-22.57925864	41.39530751		-8.095E-14	-3.574586918	3.731843007		
Count		5	11	11		5	5	5		5	5	6	6	5	5
Total		276.73	583.5049734	41.9	530.9	234.4009036	12.56286574	234.4009036	274.8179886	-2.57474E-13	2.000691891	-4.970353517	9.65860284	1.069461649	0.281815433
Mean		55.346	53.04590667	3.809	48.26	46.88018072	2.093810957	46.88018072	45.8029981	5.14947E-14	0.400138378	-0.828392253	1.60976714	0.21389233	0.056363087
Pop.SD		13.4279	9.420334107	21.54	9.731	3.631104032	21.41393096	3.830758561	3.830758561	1.64508E-13	1.514341329	1.297201344	1.5875/0025	0.219771682	0.031862541
Σ(Q)²		276.73	259.2154817	34.7	236.2	35.14212438	5.85702073	35.14212438	233.426811	-2.57474E-13	2.000691891	-1.3957726	5.926759833	1.069461649	0.281815433
Σ(p-q)²/count		55.346	23.56504379	3.155	21.47	7.028424877	5.85702073	46.88018072	38.90378019	-5.14947E-14	0.400138378	-0.232628767	0.987793305	0.21389233	0.056363087
RMSE		7.439489	4.854383977	1.776	4.634	2.651117666	2.420128247	6.846910305	6.23728949	2.26924E-07	0.632256492	0.482316044	0.993877913	0.462484951	0.237409112

Table 5.2
The Precision of Points
(Ideal Test Ellipse)

		A	B	C	D	E	F	G	H	I	J	K	L	M	N	O	P	Q	R

Col			RADIAL LINE LENGTH		POINTS		Computed (EXCEL)		RADIUS	RADIUS	PSDs OF RADII	PSDs OF POINTS		PSD	RADIAL LINE
		PLOT	COMPUTED	Plot				Plot	Excel	EXCEL vs. Plot	Plot vs. EXCEL		Of	SQUARED ERROR FRACTION
Serial	Point	From Origin Zero	From Origin Zero	x	y	x	y				x	y	Plot Lines	
---	---	---	---	---	---	---	---	---	---	---	---	---	---	---
1	p1	62.62	62.64982043	-30	55			62.64982043					-0.047621256	0.055293754
2	p2	40	39.90050125	-6.3	39.4			39.90050125					0.248746875	0.065481446
3	p3	76.85	76.4852927	15	75			76.4852927					0.47470327	0.099143764
4	p4	43.13	43.14313387	17.8	39.3			43.14313387					-0.030451817	0.061524401
5	p5	54.13	53.90037106	32.5	43			53.90037106					0.42421752	0.0003372068
6	q1		47.91795071	-22.7	42.2	-22.57925864	41.39530751	47.91795071	47.15288331	-1.622525167	0.531900264	1.906854249		
7	q2		50.11077728	-7.8	49.5	-7.844253334	49.67838522	50.11077728	50.29387903	0.364063682	-0.567350437	-0.360374174		
8	q3		52.40229003	10.2	51.4	10.22052134	51.10260668	52.40229003	52.11463773	-0.551960652	-0.20118957	0.57858624		
9	q4		53.84440547	22.1	49.1	22.17415714	47.35865283	53.84440547	52.29278385	-2.967181131	-0.335552674	3.546531913		
10	q5		54.94005825	32.9	44	33.17095788	43.88772889	54.94005825	55.01313655	0.13283792	-0.823580182	0.25516162		
11	t1		48.21037233	-21.8	43	-22.57925864	41.39530751	48.21037233	47.15288331	-2.242681561	-3.574580918	3.731843007		
Count		5	11	11	11	6	6	11	6	6	6	6	5	5
Total		276.73	583.5049734	41.9	530.9	12.56286574	274.8179886	583.5049734	304.0202038	-6.887446905	-4.970353517	9.65860284	1.069461649	0.281815433
Mean		55.346	53.04590667	3.809	48.26	2.093810957	45.8029981	53.04590667	50.67003396	-1.147907817	-0.828392253	1.60976714	0.21389233	0.056363087
Pop.SD		13.4279	9.420334107	21.54	9.731	21.41393096	3.830758561	9.420334107	2.842092168	1.225769626	1.297201344	1.58750025	0.219771682	0.031862541
$\Sigma(Q)^2$								259.2154817	256.8673205	-4.644765344	-1.3957726	5.926759833	1.069461649	0.281815433
$\Sigma(p-q)^2$/count								23.56504379	42.81122008	-0.774127557	-0.232628767	0.987793305	0.21389233	0.056363087
RMSE								4.854383977	6.543028357	0.879845189	0.482316044	0.993877913	0.462484951	0.237409112

Table 5.3
The Precision of Radii
(Ideal Test Ellipse)

Table 5.4 — The Precision of Deviations (Ideal Test Ellipse)

Serial	Point	PLOT From Origin Zero	COMPUTED From Origin Zero	Plot x	Plot y	Computed (EXCEL) x	Computed (EXCEL) y	RADIUS Plot	RADIUS Excel	DEVIATION δ Plot	DEVIATION δ Excel	PSDs OF DEVIATIONS EXCEL vs. Plot	PSDs OF RADII EXCEL vs. Plot	PSDs OF POINTS Plot vs. EXCEL x	PSDs OF POINTS Plot vs. EXCEL y	PSD Of Plot Lines	RADIAL LINE SQUARED ERROR FRACTION
1	p1	63.62	62.64982043	-30	55			62.64982043		14.73533169						-0.047621256	0.053793754
2	p2	40	39.90050125	-6.3	39.4			39.90050125		10.21077862						0.248746875	0.065481446
3	p3	76.85	76.4852927	15	75			76.4852927		24.08518916						0.474570327	0.099143764
4	p4	43.13	43.14313387	17.8	39.3			43.14313387		10.701869						-0.030051817	0.061524401
5	p5	54.13	53.90037106	32.5	43			53.90037106		1.070037961						0.42421752	0.000372068
6	q1		47.91795071	-22.7	42.2	-22.57925864	41.395730751	47.91795071	47.15288331		15.49693712	4.914565476	-1.623525167	0.531900264	1.906854249		
7	q2		50.11077728	-7.8	49.5	-7.844255334	49.67838522	50.11077728	50.29387903		10.39374432	1.760344459	0.364063682	-0.567350437	-0.360374174		
8	q3		52.40259003	10.2	51.4	10.22052134	51.10260668	52.40259003	52.11463773		24.37065497	1.179557195	0.553960652	-0.20118957	0.578686224		
9	q4		53.84440547	22.1	49.1	22.17415714	47.35865283	53.84440547	52.29278385		9.169249486	-16.71477597	-2.567181131	-0.335552674	3.546531913		
10	q5		54.94005825	32.9	44	33.17095788	41.86772689	54.94005825	55.01313655		1.117765498	3.211147058	0.132837925	-0.823680182	0.25516162		
11	t1		48.21037233	-21.8	43	-22.57925864	41.395730751	48.21037233	47.15288331				-2.426811561	-3.574580918	3.731843007		
Count		5	11	11		6	6	11	6	5	5		6	6	6	5	5
Total		276.73	583.5049734	41.9	530.9	12.56286574	274.8179886	583.5049734	304.0202038	60.80820142	60.5433514	-5.649117678	6.887446905	-4.970353517	9.65860284	1.069461649	0.281815433
Mean		55.346	53.04590667	3.809	48.26	2.093810957	45.80299681	53.04590667	50.67003396	12.16164028	12.10867028	-1.129834536	-1.149790781	-0.828992253	1.609976714	0.21389733	0.056363087
Pop.SD		13.4279	9.420334107	21.54	9.731	21.41393096	9.420334107	2.842092168	7.450537268	7.673388264	7.898804784	1.225769626	1.297201344	1.58757000025	0.219771682	0.031862541
Σ(Q)²								259.21546817	756.8673205	60.80820142	60.5433514	-5.649117678	-4.644765344	-1.3957726	5.926759833	1.069461649	0.281815433
Σ(p-q)²/count								23.56504379	42.83122008	12.16164028	12.10867028	-1.129834536	-0.774127557	-0.232628767	0.987793305	0.21389233	0.056363087
RMSE								4.854385977	6.543028357	3.487354339	3.479751468	1.06293675	0.879845189	0.482316044	0.993877913	0.462484951	0.237409112

Table 5.4
The Precision of Deviations
(Ideal Test Ellipse)

<u>Criticism</u>

A   The Deviations $\delta_i = R_i - S_i$ are not Orthogonal to the Ellipse Perimeter

Because the point rays are referred to the general origin at O and not the ellipse center at J (or indeed shifted K) no ray is at right angles to the ellipse except at angles 0, $\pi/2$ and π.

Reference to Figure 5.3 confirms that the ray p1 to t1 orthogonal to the ellipse trace does not intersect the origin O. It is, however, clear that as the deviation of the point from the ellipse decreases the line segments $p_i - q_i$ and $p_i - t_i$ become ever more equal, without of course matching exactly.

My opinion is that *the principle* of stone circuit construction as a series of offset ellipses can be established and compared with other constructs without the more complex and fractured analytic geometry necessitated by trace orthogonality.

B   Any Fitted Ellipse has Insufficient Degrees of Freedom

It is my intent to model the practice of geoscription as it might have been applied by early humans. Therefore, we allow them to vary Loop Length, L, and make it suit a system of four stakes, two of which define the foci of an ellipse in any given quarter.

Accordingly, A and B, the Ellipse Axes may also be defined at will, whilst the Interfocal Distance, 2c, is implicitly defined by the separation of the two active turnposts.

It is admittedly true that the rotation of the unshifted and shifted ellipses is not possible in such a simple system, and therefore I have discounted rotation of the traces.

C   Figure Fitting by Hand and Eye is Unnecessarily Crude and Imprecise

The literature contains several schemes for the optimal (usually Least Squares) fitment of ellipses to point clouds and almost equally frequently algorithms are published therewith. The principle applied is usually the Solution of Sets of Non-Linear equations.

In my opinion, these methods are unnecessarily complicated and potentially expensive for the quality and amount of data available from Boscawen-Ûn, and almost always beyond the intellectual powers of this researcher.

As I remarked above, the archaeological *principle* can be proven to the satisfaction of reasonable standards using manual, so to say, craft mathematics.

D   The Natural System is Vitiated by "Land Creep"

I accept that the site of the Boscawen-Ûn monument has warped and distorted during the four or six millennia since its construction.

This is made visible by the fact that post-and-rope geometry does not fit simple systems of geoscription with exactitude.

In the wet British climate there is a tendency for topsoil and underlying rock gradually to slump shoreward under gravity in a process of solifluction intensified by freeze-thaw effects.

At least one circuit stone has almost certainly been removed to an unknown place sometime since construction. The rest of the megaliths have been shifted individually and collectively by vegetation roots, animal burrowing and other vagaries of existence.

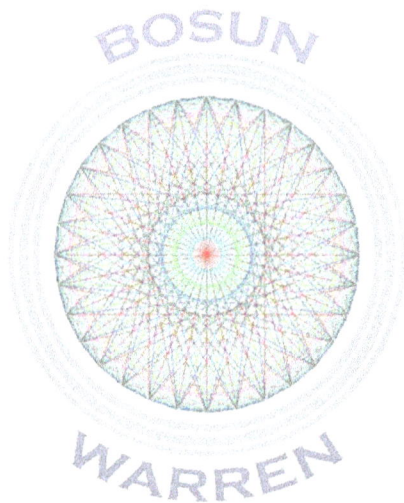

CHAPTER SIX
ROTATIVE FITMENT

The essence of rotative fitment is that each of the four sectorial sets of stone points are rotated to "top center" as a convenience to analysis and illustration. Like all "conveniences" this is controversial, not least because it is a potential avenue of adventitious error introduction.

The required equations for the Rigid Rotation of a Point about a Center are:-

Sector	px_i	py_i
1	$-x_{bos}Cos(\alpha)$	$-y_{bos}Cos(\alpha)$
2	$-y_{bos}Sin(\alpha)$	$+x_{bos}Sin(\alpha)$
3	$-x_{bos}Cos(\alpha)$	$-y_{bos}Cos(\alpha)$
4	$-y_{bos}Sin(\alpha)$	$+x_{bos}Sin(\alpha)$

Table 6.1
The Formulae for the Rotation of Points

The formulae of Table 6.1 assume that rotation is about the Origin at (0,0), in our terms the barycenter.

The Counter Rotation Constant, α, is an angle given by:-

$$\alpha = -\frac{\pi}{2}.(S-1)$$
Equation 6.1

where S is the Sector Serial Number {1,2,3,4}.

Once the Points (px_i,py_i) are rotated to the Northern "top center" zone, the relevant rotated (qx_i,qy_i) can be established using:-

$$qx_i$$

$$= \frac{b^2.h_e - a^2.g_i^2.\varphi_i + Sgn_i.a.b.\sqrt{b^2 + a^2.g_i^2 - 2.g_i.\varphi_i.h_e - \varphi_i^2 - g_i^2.h_e^2}}{b^2 + a^2.g_i^2}$$

$$\times \frac{1}{C_S}$$

Equation 5.22

$$qy_i = \frac{b^2.\mu_i + a^2.g_i^2.k_e + Sgn_i.a.b.g_i.\sqrt{b^2 + a^2.g_i^2 + 2.\mu_i.k_e - k_e^2 - \mu_i^2}}{b^2 + a^2.g_i^2}$$

$$\times \frac{1}{C_S}$$

Equation 5.23

in a usual manner except that, as shown, the Equations 5.22 and 5.23 are adjusted by being divided by the Ellipse Scaling Factor, C_S.

The Sign Sgn_i is given by:-

$$Sgn_i = Sign(px_i)Sgn_i = Sign(px_i)$$

Equation 6.2

Remember that although bos goes from 1 to 19, the Number of Circuit Stones j never exceeds five, the Number of Stones in a Given Sector.

It behoves us to bear this in mind if we compute both Sectoral (j) and Aggregate (i) statistics.

Descriptive Statistics

We have (px_i, py_i) aboriginally, so to say, because they are nothing other than the RRZSI stone Position Points (x_{bos}, y_{bos}). By definition the (barycentric) Origin is (0,0).

Therefore, the straight-line Point Radius, R_i, is given by:-

$$R_i = \sqrt{(px_i - 0)^2 + (py_i - 0)^2} = \sqrt{(px_i)^2 + (py_i)^2}$$

Equation 6.3

and the co-linear Ellipse Intersection Radius, r_i, is given by:-

$$r_i = \sqrt{(qx_i - 0)^2 + (qy_i - 0)^2} = \sqrt{(qx_i)^2 + (qy_i)^2}$$
Equation 6.4

Accordingly, the Deviations, δ_i, are given by:-

$$\delta_i = R_i - r_i$$
Equation 6.5

In these terms the Mean Standard Error, MSE, is:-

$$MSE = \mu[\delta_i^2] = \frac{\sum \delta^2}{n}$$
Equation 6.6

where m is the Mean of the bracketed metrics and n is the Number of Circuit Stones, i.e. 19.
Also, the Root Mean Square, RMS (error), is:-

$$RMS = \sqrt{MSE}$$
Equation 6.7

and Dimensionless MSE ε'_μ is :-

$$\varepsilon'_\mu = \frac{\sum \delta_i^2}{\sum R_i^2}$$
Equation 6.8

Stone Serial				Serial by Sector		
	xbos	**ybos**			**xbos**	**ybos**
1	-13.200887	-8.578189		12	6.5827515	11.454973
2	-10.023523	-11.360169		13	3.2273643	12.865783
3	-5.6331766	-13.351117		14	-1.1366474	13.4525948
4	-0.6890507	-13.858837		15	-5.5270545	12.958289
5	3.78704709	-13.674366		16	-9.7592682	11.6861015
6	8.79713309	-12.626331		17	-13.543255	8.71971153
7	12.4888351	-9.4292128		18	-15.138673	4.29639497
8	14.8225647	-5.5135126		19	-15.725489	-0.2587901
9	15.1984349	-0.8132943		1	-13.200887	-8.578189
10	14.0053816	4.5593654		2	-10.023523	-11.360169
11	11.4675114	9.47060529		3	-5.6331766	-13.351117
12	6.5827515	11.454973		4	-0.6890507	-13.858837
13	3.2273643	12.865783		5	3.78704709	-13.674366
14	-1.1366474	13.4525948		6	8.79713309	-12.626331
15	-5.5270545	12.958289		7	12.4888351	-9.4292128
16	-9.7592682	11.6861015		8	14.8225647	-5.5135126
17	-13.543255	8.71971153		9	15.1984349	-0.8132943
18	-15.138673	4.29639497		10	14.0053816	4.5593654
19	-15.725489	-0.2587901		11	11.4675114	9.47060529
Σ	0	-1.199E-13	Σ		-1.599E-14	-1.35E-13
μ	0	-6.311E-15	μ		-8.414E-16	-7.105E-15
σ	10.6960933	10.3277351	σ		10.6960933	10.3277351

Table 6.2
The Formulae for the Rotation of Points

Sector	Counter-Rotated Data		Counter-Rotated Intercepts	
	px	py	qx	qy
1	6.582752	11.45497	7.334656	12.7634
1	3.227364	12.86578	3.379245	13.47125
1	-1.13665	13.45259	-1.152009	13.63441
1	-5.52705	12.95829	-5.605378	13.14192
1	-9.75927	11.6861	-9.980556	11.95108
2	8.719712	13.54326	9.24722	14.36257
2	4.296395	15.13867	4.224069	14.88383
2	-0.25879	15.72549	-0.24215	14.71432
2	-8.57819	13.20089	-8.301879	12.77568
3	10.02352	11.36017	10.08286	11.42741
3	5.633177	13.35112	5.612671	13.30252
3	0.689051	13.85884	0.696981	14.01833
3	-3.78705	13.67437	-3.795192	13.70378
3	-8.79713	12.62633	-8.535153	12.25032
4	9.429213	12.48884	11.3183	14.9909
4	5.513513	14.82256	5.799498	15.59141
4	0.813294	15.19843	0.823396	15.38722
4	-4.55937	14.00538	-4.657486	14.30679
4	-9.47061	11.46751	-10.03441	12.15019
Σ	3.053892	252.8796	6.214681	258.8273
μ	0.160731	13.30945	0.327088	13.62249
σ	6.502413	1.272711	6.780692	1.176519

Table 6.3
The Counter-Rotated Data and the Counter-Rotated Intercepts

	Radius Ray Length to Stone Point	Radius Ray Length to Fitted Ellipse Intersection	Deviation $R_i - r_1$	Squared Deviation $(R_i - r_i)^2$	Percentage Specific Defect	
	R_i	r_i			$PSD(R_i, r_i)$	$(R_i)^2$
	13.21170026	14.72078634	-1.509086082	2.277340804	-11.42234575	174.5490237
	13.26439795	13.88862388	-0.624225926	0.389658007	-4.70602532	175.9442531
	13.50052866	13.68298738	-0.182458719	0.033291184	-1.351493145	182.2642742
	14.08778137	14.28741796	-0.199636593	0.039854769	-1.417090371	198.4655839
	15.22525151	15.57047839	-0.345226883	0.119181601	-2.267462595	231.8082834
	16.10754877	17.08199175	-0.974442981	0.949539123	-6.049604411	259.4531274
	15.73653127	15.47162158	0.264909691	0.070177145	1.683405872	247.6384163
	15.72761830	14.71631083	1.011307477	1.022742814	6.430137467	247.3579775
	15.74321268	15.23611171	0.507100969	0.257151393	3.221076788	247.8487454
	15.15006419	15.23974379	-0.089679593	0.008042429	-0.591941997	229.5244451
	14.49085914	14.43811106	0.052748074	0.002782359	0.364009296	209.9849985
	13.87595559	14.03564604	-0.159690451	0.02550104	-1.150842908	192.5421435
	14.18908029	14.21959820	-0.030517912	0.000931343	-0.215080271	201.3299994
	15.38875555	14.93047594	0.458279616	0.210020206	2.978016087	236.8137974
	15.64867584	18.78379408	-3.135118235	9.82896635	-20.0343995	244.8810557
	15.81477936	16.63509113	-0.82031177	0.6729114	-5.18699472	250.1072461
	15.22017968	15.40923290	-0.18905322	0.03574112	-1.242122127	231.6538696
	14.72883320	15.04580692	-0.316973725	0.100472342	-2.152062699	216.9385273
	14.87266561	15.75806324	-0.885397627	0.783928957	-5.953187206	221.1961825
Σ	281.98441922	289.15189311	-7.167473889	16.82823439	-49.06400751	4200.30195
μ	14.84128522	15.21852069	-0.377235468	0.885696547	-2.582316185	221.0685237
σ	0.89709344	1.19909646	0.862200643	2.178859232	5.650647133	26.26279411

MSE		0.885696547
ε'_μ		0.004006434
RMS		0.941114524

Table 6.4
Radius Ray and Deviational Results

Stone Serial for SORT	xbos	ybos	Swept Angle of Stone about Origin	Radius Ray Length to Stone Point	Radius Ray Length to Fitted Ellipse Intersection
			ω	R_i	r_i
1	-13.200887	-8.578189	-2.56534	15.74321	15.2361117
2	-10.023523	-11.360169	-2.29377	15.15006	15.2397438
3	-5.6331766	-13.351117	-1.97006	14.49086	14.4381111
4	-0.6890507	-13.858837	-1.62047	13.87596	14.035646
5	3.78704709	-13.674366	-1.30062	14.18908	14.2195982
6	8.79713309	-12.626331	-0.96227	15.38876	14.9304759
7	12.4888351	-9.4292128	-0.6467	15.64868	18.7837941
8	14.8225647	-5.5135126	-0.35611	15.81478	16.6350911
9	15.1984349	-0.8132943	-0.05346	15.22018	15.4092329
10	14.0053816	4.5593654	0.314724	14.72883	15.0458069
11	11.4675114	9.47060529	0.690314	14.87267	15.7580632
12	6.5827515	11.454973	1.049215	13.2117	14.7207863
13	3.2273643	12.865783	1.325019	13.2644	13.8886239
14	-1.1366474	13.4525948	1.655089	13.50053	13.6829874
15	-5.5270545	12.958289	1.973959	14.08778	14.287418
16	-9.7592682	11.6861015	2.266587	15.22525	15.5704784
17	-13.543255	8.71971153	2.569559	16.10755	17.0819918
18	-15.138673	4.29639497	2.865061	15.73653	15.4716216
19	-15.725489	-0.2587901	3.158048	15.72762	14.7163108
Σ			6.098765	281.9844	289.151893
μ	6.28318531		0.320988	14.84129	15.2185207
σ			1.770327	0.897093	1.19909646

Table 6.5
Sorted Interangular Results for Stones in
Original (not Sectoral) Order

<u>Tabulations</u>

For the sector-wise fitment of ellipses it is "convenient" to re-arrange the RRSZI stone position data to group the stones by sector, without compromising intra-sector sequentiality. The Original and Arranged stone co-ordinates are tabulated in Table 6.2:-

On this basis, the Counter-Rotated Data and the Counter-Rotated Intercepts; Deviational; and Interangular Results are given by Tables 6.2 through 6.4

<u>Tables that Relate to the Fitment of Ellipses to Sector Stone Circuit Data</u>

All ellipse fitments were done by hand and eye. No attempt was made at the use of Non-Linear Simultaneous Equation Solutions, whether Least Squares or otherwise.

Table 6.6

Fitted Sector Ellipse Scaling, Shift and Rotation Parameters:
Fitted Sector Ellipse Axes and Loop Parameters

Sector	Cardinal	C_s	xs	ys	xP	yP	yP	α_analytic	α	α
1	NORTH	0.65	2	4.25	0	8.876215213	0 B/2	0	0	0
2	WEST	0.6	1.8	0	0	9.277177165	0 A/2	-1.5708	-1.5708	$-(\pi/2)$
3	SOUTH	0.65	0	3.75	0	8.876215213	0 B/2	-3.14159	-3.14159	$-(\pi)$
4	EAST	0.6	2.5	0	0	9.277177165	0 A/2	-4.71239	-4.71239	$-(3\pi/2)$
	Means	0.625	1.575	2	0	9.076696189	0	-2.35619	-2.35619	
	Pop.SDs	0.025	0.944391338	2.007797301	0	0.200480976	0	1.756204	1.756204	

Sector	Cardinal	A	B	D	a	b	c	L
1	NORTH	20.10055052	17.75243043	26.81754868	13.40877434	8.87621521	10.05027526	82.42296006
2	WEST	18.55435433	16.38685885	24.75466032	12.37733016	8.19342943	9.27717716	76.08273236
3	SOUTH	20.10055052	17.75243043	26.81754868	13.40877434	8.87621521	10.05027526	82.42296006
4	EAST	18.55435433	16.38685885	24.75466032	12.37733016	8.19342943	9.27717716	76.08273236
	Means	19.32745243	17.06964464	25.7861045	12.89305225	8.53482232	9.663726213	79.25284621
	Pop.SDs	0.773098097	0.682785786	1.03144418	0.51577209	0.341392893	0.386549049	3.170113848

Table 6.7 is a synopsis of relevant Sector Data:-

SECTOR NUMBER		1	2	3	4
SECTOR NAME		NORTH	WEST	SOUTH	EAST
ROTATION CONSTANT	α	0	-1.570796	-3.141593	-4.712389
Scaling Modifier	S	0.65	0.6	0.65	0.6
x-Shift	xs	2	1.8	0	2.5
y-Shift	ys	4.25	0	3.75	0
Rectangle Width	A	20.100551	18.554354	20.100551	18.554354
Rectangle Height	B	17.752430	16.386859	17.752430	16.386859
Rectangular Diagonal	D	26.817549	24.754660	26.817549	24.754660
Ellipse Major Semi-Axis (Width)	a	13.408774	12.377330	13.408774	12.377330
Ellipse Minor Semi-Axis (Height)	b	8.876215	8.193429	8.876215	8.193429
Ellipse Focal Distance	c	10.050275	9.277177	10.050275	9.277177
Loop Length	L	82.422960	76.082732	82.422960	76.082732
Half-Ellipse Lower Bound	Lbe	0	0	0	0
Half-Ellipse Upper Bound	Ube	3.141593	3.141593	3.141593	3.141593
Number of Half-Ellipse Increments	NINCe	32	32	32	32
Half-Ellipse Increment	INCe	0.098175	0.098175	0.098175	0.098175
Ordinal Subject Axis (Sector X)	X = B	17.752430	18.554354	17.752430	18.554354

Table 6.7
Fitted Sector Ellipse Axes and Loop Parameters
Graphical Illustrations

In terms of the mental conception of the graphics it is important to remember that we have rotated all Sectors to the Northern Sector native to Sector One (Red) in order to "facilitate" analysis.

A corollary of this is that whilst the fitted ellipse for Sector One fits the Sector One stones nicely, the other three sectors do not seem to fit them. This is a conceptual or cognitative artefact. The other sectors fit their own stones more or less neatly.

Note especially that the gray trace and triangle denoting the stone circuit trace are included for *contextual illustration only*, and in no way stand comparison with colored fitment traces, except in the case of Sector One only.

Figure 6.1 shows the circuit trace for background only whilst being a composite of the four sector-wise ellipse-arc fitments.

The linear gray dashed traces radiating from the center illustrate selected pq0 co-linearities.

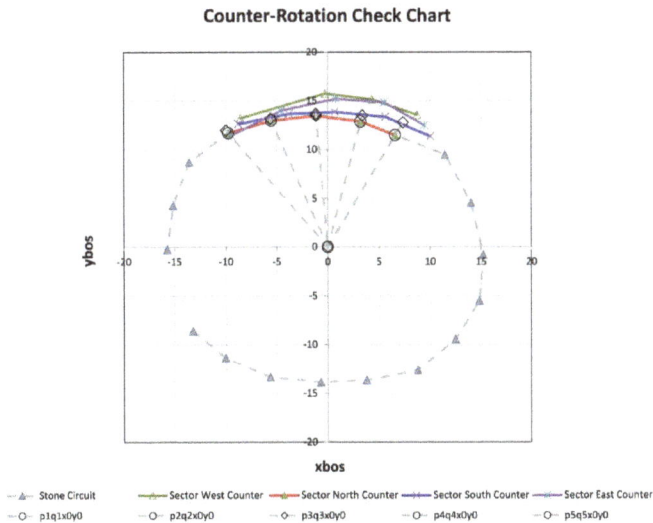

Counter-Rotation Check Chart

Figure 6.1
Schematic Composite of Fitted Sectoral Traces

Note that it fallaciously appears from Figure 6.1 that only the Sector One ellipse segment precisely fits the Northern stones.

To appreciate the excellent fit of *all* the elliptic traces please study Figure 6.2 which presents the *unrotated* fitments:-

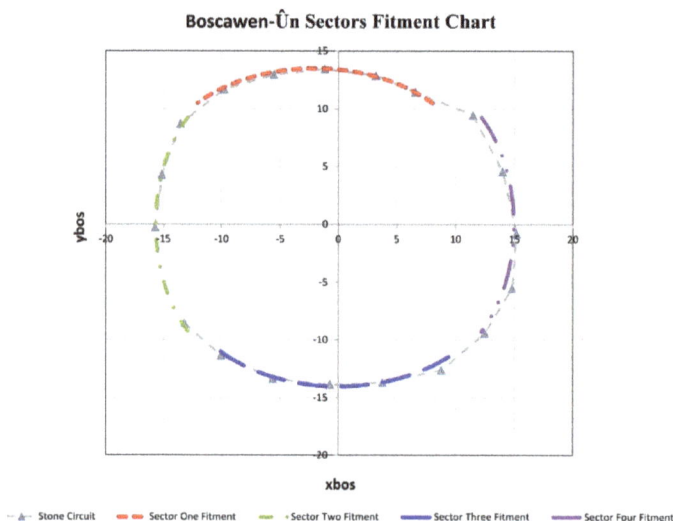

Figure 6.2
Fitted Sectoral Traces for the Boscawen-Ûn Stone Circuit

It can now been seen that the ellipse fit, and therefore the implication of focal turnposts, is excellent.

Remember that these ellipse arcs were fitted by a human craftsman by hand and eye and that scientifically-optimised fitments using non-linear matrix optimisation or something would yield even more convincing correlations.

Even the existing fitments could be improved by a younger man with better eyesight and a steadier right hand.

CHAPTER SEVEN

NON-ROTATIVE FITMENTS

In our previous discussion we rotated three of the sectors to assist constructive analyses.

In this Chapter we shall study the advantages to be accrued, if any, from a direct approach working with mathematically undisturbed stone points.

Figure 7.1 is a Veusz® plot that shows the stones with four new ellipse fitments:-

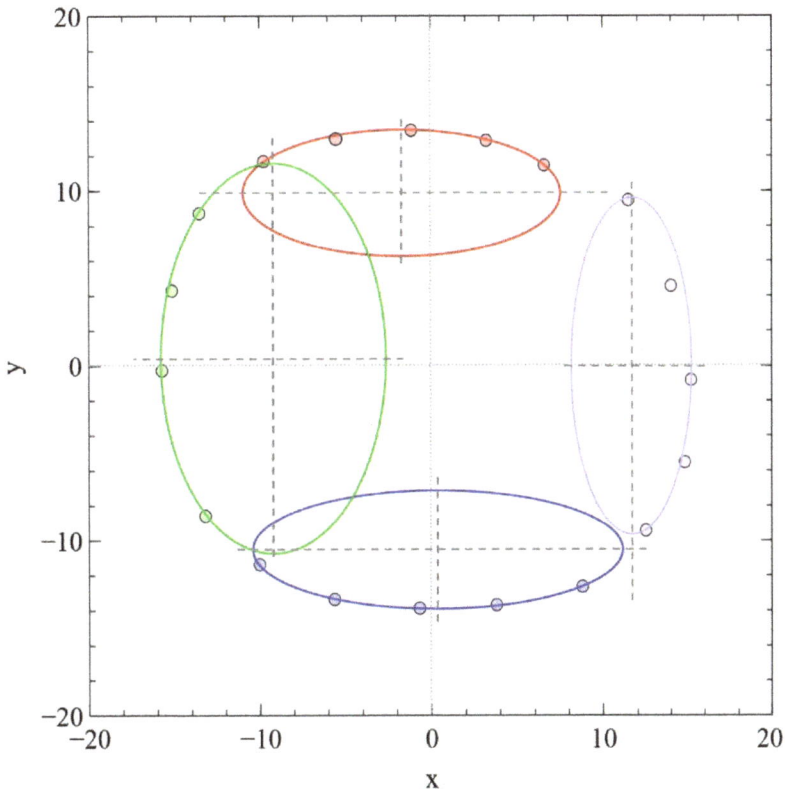

Figure 7.1
The Boscawen-Ûn Stones with
New Sector Ellipse Fitments

The new fittings are hand-and-eye fittings.

The usual color coding is red for Sector One (North); green for Sector Two (West); blue for Sector Three (South) and purple for Sector Four (East). The (complete) ellipses are hand-and-eye fits to the circuit stones mapped to RRZSI ground position by the centers of the fimbriated colored circles. The gray pecked lines denote the A and B axes of the respective ellipses, and the intersections of those lines are the respective ellipse centers. The calibration is in ground meters.

Tabulations and their Algebraic Bases: Active Calibration Forms

Table 7.1 is a representation of the Photodraw® point co-ordinate readings as against their equivalent Veusz® graph positions which tally directly with Ground co-ordinates (in actual meters). The forward-substantive outcome is an x-Abscissal Calibration Mean of 0.316330615 and a y-Ordinal Calibration Mean of 0.317082888.

This implies that that there are about 0.3165 ground meters to the PhotoDraw millimeter.

The slight discrepancy in the averages is due to my inability to scale the plot to exact equiaxial symmetry.

The relevant equation is:-

$$C_{VPd} = \frac{NWX_{Veusz} - NEX_{Veusz}}{NWX_{PhotoDraw} - NEX_{PhotoDraw}}$$

Equation 7.1

Analogous calculations for the other three plot borders are made and the respective horizontal and vertical dividends averaged.

Table 7.2 extends the calibration process to reconcile the plot corner and fitted ellipse readings with the equivalent ground displacements in order to facilitate real-life point and ellipse positions and dimensions.

PhotoDraw is a product of the last century and in conformance with computer programming practice of the time measures displacements from the *top* of the figure.

Accordingly, I had to invert the ordinal displacements *only* in order to conform vertical ground displacements with the Cartesian bottom-up convention.

Table 7.3 presents the relevant figures that exit this process.

Table 7.4 presents the analogous computations as applied to the hand-fitted sector ellipses: First ordinal inversion; then calibration to Veusz/Ground scalings. Further ellipse-related work is needed to produce the actual ellipse parameters as quoted in Table 7.5

Table 7.6 presents the resulting fitted Ellipse Parameters and their Implied Foci as they occur at their Cornish site of Boscawen-Ûn.

CALIBRATION ORIGIN

	x	y
Base Position (meters)	0	0
PhotoDraw Position (mm)	84.7	73.4
Implied Origin	26.7932031	25.4934642

CORNER

	NE		NW		SW		SE	
	x	y	x	y	x	y	x	y
Base Position (meters)	20	20	20	-20	20	-20	-20	20
PhotoDraw Position (mm)	10.3	147.9	10.3	21.4	10.3	21.4	136.5	147.8
							-20	136.4

Meters Per mm

NW-NE	0.31620553
SW-SE	0.3164557
NW-SW	0.31695721
NE-SE	0.31720856
CALIBRANT	0.31670675
Oblation	0.99762752
PhotoDraw y-Height	153.8
VeuszBase y-Height	48.7094984
CALIBRATION MEAN	0.31633061
	0.31708289

Table 7.1
Boscawen-Ûn
VeuszBase Calibration Form

OBJECTS (Inverted Ex PhotoDraw)

Object	PhotoDraw Left	Top	Width	Height	PhotoDraw Position Left	Right	Measured Center x	y	Computed Center x	y	Ground Map Left	Top	Width	Height	Computed Center x	y
Origin	84.6	73.3	0	0			84.6	73.3	84.6	73.3	20.01587	19.95253	NA	NA	26.79339	23.2146
Horizontal Origin Line	21.46	73.41	126.34						84.63	73.41	0.019002	19.98736	40.01273	0		
Vertical Origin Line	84.62	10.12	126.34						147.79	73.29	20.0222	-0.05701	40.01273	0		
Veusz Graph Square	21.4	10.4	126.5	125.9			84.5	73.3	84.65	73.35	0	0.031671	40.0634	39.87338	84.65	73.35
North Ellipse	50.04	31.07	58.55	22.78			79.2	42.3	79.315	42.46	9.070481	6.577999	18.54318	7.21458	18.34207	10.18529
West Ellipse	34.82	37.08	41.63	70.29			55.4	72.1	55.635	72.225	4.250205	8.481407	13.1845	22.26132	10.84246	19.61207
South Ellipse	51.82	96.01	68.14	21.34			85.9	106.5	85.89	106.68	9.634219	27.14494	21.5804	6.758522	20.42442	30.5242
East Ellipse	110.55	43.24	22.35	60.71			121.6	73.6	121.725	73.595	28.23441	10.43232	7.078396	19.22727	31.7736	20.04595

OBJECTS (Corrected y's)

Object	PhotoDraw Left	Top	Width	Height	PhotoDraw Position Left	Right	Measured Center x	y	Computed Center x	y	Ground Map Left	Top	Width	Height	Computed Center x	y
Origin	84.6	73.3	0	0			84.6	73.3	84.6	73.3	20.01587	19.95253	NA	NA	11.89712	9.814711
Horizontal Origin Line	21.46	73.41	126.34						84.63	73.41	0.019002	19.98736	40.01273	0		
Vertical Origin Line	84.62	10.12	126.34						147.79	73.29	20.0222	-0.05701	40.01273	0		
Veusz Graph Square	21.4	10.4	126.5	125.9			84.5	73.3	84.65	73.35	0	0.031671	40.0634	39.87338	84.65	73.35
North Ellipse	50.04	122.7	58.55	22.78			79.2	42.3	79.315	111.34	9.070481	35.60734	18.54318	7.21458	18.34207	22.57327
West Ellipse	34.82	116.7	41.63	70.29			55.4	72.1	55.635	81.575	4.250205	33.70393	13.1845	22.26132	10.84246	11.66114
South Ellipse	51.82	57.79	68.14	21.34			85.9	106.5	85.89	47.12	9.634219	15.0404	21.5804	6.758522	20.42442	30.5242
East Ellipse	110.55	110.6	22.35	60.71			121.6	73.6	121.725	80.205	28.23441	31.75302	7.078396	19.22727	31.7736	22.13939

Table 7.2
Boscawen-Ûn
VeuszBase Objects Form

| Plot | MEASURED | | | | | | GROUND | | |
Corner	x	y	y_{inv}	$x_{shifted}$	$y_{shifted}$	y_{inv}	x	y	y_{inv}
SW Photodraw Base	0	153.9	-0.1	-21.4	143.4	10.4	-6.77752	45.41575	3.2937502
NE	147.8	10.5	143.3	126.4	0	153.8	40.03173	0	48.709498
NW	21.4	10.5	143.3	0	0	153.8	0	0	48.709498
SW	21.4	136.3	17.5	0	125.8	28	0	39.84171	8.867789
SE	147.8	136.4	17.4	126.4	125.9	27.9	40.03173	39.87338	8.8361184
CENTER	84.6	73.2	80.6	63.2	62.7	91.1	20.01587	19.85751	28.851985
Calibrant	0.316706751								

Table 7.3
Boscawen-Ûn
Plot Corners Checks Form

MEASURED

Fitted Ellipse	Center		West		South		East		North	
	x	y	x	y	x	y	x	y	x	y
NORTH	79.2	42.3	50	42.3	79.2	53.8	108.6	42.2	79.3	30.9
WEST	55.5	72.1	34.7	72.2	55.5	107.3	76.3	72.1	55.5	37
SOUTH	85.9	106.6	51.9	106.7	85.8	117.3	119.9	106.6	85.9	96
EAST	121.6	73.6	110.5	73.5	121.6	103.8	132.8	73.6	121.7	43.1

COMPUTED (y-inversion)

Fitted Ellipse	Center		West		South		East		North	
	x	y	x	y	x	y	x	y	x	y
NORTH	-5.4	31	-34.6	31	-5.4	19.5	24	31.1	-5.3	42.4
WEST	-29.1	1.2	-49.9	1.1	-29.1	-34	-8.3	1.2	-29.1	36.3
SOUTH	1.3	-33.3	-32.7	-33.4	1.3	-44	35.3	-33.3	1.3	-22.7
EAST	37	-0.3	25.9	-0.2	37	-30.5	48.2	-0.3	37.1	30.2

GROUND

Fitted Ellipse	Center		West		South		East		North	
	x	y	x	y	x	y	x	y	x	y
NORTH	-1.70819	9.82957	-10.945	9.82957	-1.70819	6.183116	7.591935	9.82957	-1.67655	13.44431
WEST	-9.20522	0.380499	-15.7849	0.348791	-9.20522	-10.7808	-2.62554	0.380499	-9.20522	11.51011
SOUTH	0.41123	-10.5589	-10.344	-10.5906	0.379597	-13.9516	11.16647	-10.5589	0.41123	-7.19778
EAST	11.70423	-0.09512	8.192963	-0.06342	11.70423	-9.67103	15.24714	-0.09512	11.73587	9.575903

Table 7.4
Boscawen-Ûn
Ellipses Checks Form

MEASURED

Fitted Ellipse	Center		West		South		East		North	
	x	y	x	y	x	y	x	y	x	y
NORTH	79.2	42.3	50	42.3	79.2	96	108.6	42.2	79.3	30.9
WEST	55.5	72.1	34.7	72.2	55.5	107.3	76.3	72.1	55.5	37
SOUTH	85.9	106.6	51.9	106.7	85.8	117.3	119.9	106.6	85.9	96
EAST	121.6	73.6	110.5	73.5	121.6	103.8	132.8	73.6	121.7	43.1

COMPUTED (y-inversion)

Fitted Ellipse	Center		West		South		East		North	
	x	y	x	y	x	y	x	y	x	y
NORTH	79.2	30.9	50	30.9	79.2	-22.8	108.6	31	79.3	42.3
WEST	55.5	1	34.7	1	55.5	-34.1	76.3	1.1	55.5	36.2
SOUTH	85.9	-33.4	51.9	-33.5	85.8	-44.1	119.9	-33.4	85.9	-22.8
EAST	121.6	-0.4	110.5	-0.3	121.6	-30.6	132.8	-0.4	121.7	30.1

GROUND

Fitted Ellipse	Center		West		South		East		North	
	x	y	x	y	x	y	x	y	x	y
NORTH	25.05338	9.797861	15.81653	9.797861	25.05338	-7.22949	34.3535	9.82957	25.08502	13.41261
WEST	17.55635	0.348791	10.97667	0.317083	17.55635	-10.8125	24.13603	0.348791	17.55635	11.4784
SOUTH	27.1728	-10.5906	16.41756	-10.6223	27.14117	-13.9834	37.92804	-10.5906	27.1728	-7.22949
EAST	38.4658	-0.12683	34.95453	-0.09512	38.4658	-9.70274	42.00871	-0.12683	38.49744	9.544195

Table 7.5
Boscawen-Ûn
Measured Ellipses Form

GROUND

Fitted Ellipse	Center x₀	Center y₀	A	B	a	b	h	k	Implied -f	Implied +f
NORTH	-1.70819	9.82957	18.53697	7.261198	9.268487	3.630599	-1.70819	9.82957	-8.52781	8.527813
WEST	-9.20522	0.380499	13.15935	22.29093	6.579677	11.14546	-9.20522	0.380499	-8.99607	8.996066
SOUTH	0.41123	-10.5589	21.51048	6.753866	10.75524	3.376933	0.41123	-10.5589	-10.2113	10.21134
EAST	11.70423	-0.09512	7.054173	19.24693	3.527086	9.623466	11.70423	-0.09512	-8.95381	8.953812

Table 7.6
Boscawen-Ûn
The Ellipse Parameters and Foci
in
Actual Ground Meters

pq0 Radii: Geometrical Outcomes and Correlations

Table 7.7 shows the measured values of the Stone Position Points $p_{1\ldots19}$ and the Ellipse Intersection points $q_{1\ldots19}$ and, for the former, the quality of check correlations in terms of Percentage Specific Defects $PSD(x_{bos},px)$ and $PSD(y_{bos},py)$. It is clear that there is less than 1% discrepancy in either x or y calculations, except for the y-value of Stone 18 (about +1.843%).

The relevant equations are:-

$$px = C_x.(x - x_0)$$
Equation 7.2

$$py = C_y.(y_0 - y)$$
Equation 7.3

where C_x and C_y are respectively the x and y Calibration Means; x and y are Measured x and y Displacements and x_0 and y_0 are the x- and y-Coordinates of the Circuit Center.

Table 7.8 presents the Ellipse Parameters A and B are respectively the Major-Axes and a and b the Minor. h is the x-Shift and k is the y-Shift.

Table 7.9 presents the Ray Gradients for p and q, respectively g and g prime, centered at the Circuit Barycenter (x_0,y_0). Gradients are computed for both the ray due to (x_{bos},y_{bos}) and also for Intersections (qx,qy) as a check upon the computation of the latter and a guarantee of the Co-Linearity p,q,0. The nineteen rays are then listed in polar form.

Table 7.10 presents the Computational Auxiliaries that combine to generate the Ray-Ellipse Intersection Points $(qx_{1\ldots19},qy_{1\ldots19})$.

As previously observed:-

$$\varphi_i = \kappa_i - k_e$$
Equation 7.4

and since the common Ordinal Intercept of all radii is by definition zero:-

$$\varphi_i = -k_e$$
Equation 7.5

Similar arguments pertain to the Auxiliary μ_i:-

$$\mu_i = \kappa_i + g_i.h_e$$
Equation 5.21

which renders to:-

$$\mu_i = g_i.h_e$$
Equation 7.6

Further recall that to establish (qx,qy) we employ:-

$$qx_i = \frac{b^2.h_e - a^2.g_i^2.\varphi_i + Sgn_i.a.b.\sqrt{b^2 + a^2.g_i^2 - 2.g_i.\varphi_i.h_e - \varphi_i^2 - g_i^2.h_e^2}}{b^2 + a^2.g_i^2}$$
Equation 5.22

$$qy_i = \frac{b^2.\mu_i + a^2.g_i^2.k_e + Sgn_i.a.b.g_i.\sqrt{b^2 + a^2.g_i^2 + 2.\mu_i.k_e - k_e^2 - \mu_i^2}}{b^2 + a^2.g_i^2}$$
Equation 5.23

To facilitate composition and to assist numerical control it is useful to make the following substitutions:-

$$qx_i = \frac{U + V}{W}$$
Equation 7.7

$$qy_i = \frac{X + Y}{W}$$
Equation 7.8

where:-

$$W = b^2 + a^2 . g_i^2$$
Equation 7.9

$$U = b^2 . h_e - a^2 . g_i^2 . \varphi_i$$
Equation 7.10

$$V = SgnX_i . a . b . \sqrt{b^2 + a^2 . g_i^2 - 2 . g_i . \varphi_i . h_e - \varphi_i^2 - g_i^2 . h_e^2}$$
Equation 7.11

$$X = b^2 . \mu_i + a^2 . g_i^2 . k_e$$
Equation 7.12

$$Y = SgnY_i . a . b . g_i . \sqrt{b^2 + a^2 . g_i^2 + 2 . \mu_i . k_e - k_e^2 - \mu_i^2}$$
Equation 7.13

I was not able to determine a function for either SgnX or SgnY, but they appear to be the same for any particular ray: I established them empirically.

Table 7.11 presents the Radius Rays R_i and r_i, and functions of those preparatory to calculation of functions R_i-r_i, R_i/r_i, $(R_i$-$r_i)^2$ and R_i^2. These functions are themselves preparatory to the computation of descriptive statistics that assess the quality of ellipse fitment to the stones data.

Again, the common origin is at $(x_0,y_0) \equiv (0,0)$ and accordingly me may write:-

$$R_i = \sqrt{(xbos_i - x_0)^2 + (ybos_i - y_0)^2} = \sqrt{xbos_i^2 + ybos_i^2}$$
Equation 7.14

$$r_i = \sqrt{(qx_i - x_0)^2 + (qy_i - y_0)^2} = \sqrt{qx_i^2 + qy_i^2}$$
Equation 7.15

The chosen statistics were:-

$$MSE = \frac{\sum_{i=1}^{n}(R_i - r_i)^2}{n}$$

Equation 7.16

$$RMSE = \sqrt{\frac{\sum_{i=1}^{n}(R_i - r_i)^2}{n}}$$

Equation 7.17

and:-

$$\varepsilon_{\mu}' = \frac{\sum_{i=1}^{n}(R_i - r_i)^2}{\sum_{i=1}^{n} R_i^2}$$

Equation 7.18

Of course I assert no claim for the mathematical independence of these statistics: Only for their Swiftian convenience.

Table 7.12 gathers the data for the plotting of the Polar Profile comparative of the (xbos,ybos) and (qx,qy) radii. The Polar Profile diagram is presented in Figure 7.2

.

							Measured p		Measured q			Computed p		Computed q		PSD (xbos,px)	PSD (ybos,py)
		Serial	xbos	ybos		Sector	x	y	x	y		px	py	qx	qy		
		1	-13.2009	-8.57819		2	42.8	100.4	43.1	100.1		-13.2226	-8.62465	-13.1472	-8.54329	-0.16463	-0.54167
		2	-10.0235	-11.3602		3	52.9	109.3	52.9	109.3		-10.0277	-11.4467	-10.0306	-11.3682	-0.04148	-0.76164
		3	-5.63318	-13.3511		3	66.8	115.6	66.8	115.5		-5.63068	-13.4443	-5.63354	-13.352	0.044232	-0.69805
		4	-0.68905	-13.8588		3	82.4	117.1	82.3	117.3		-0.69593	-13.9199	-0.69199	-13.918	-0.99798	-0.44089
		5	3.787047	-13.6744		3	96.6	116.4	96.6	116.7		3.795967	-13.698	3.81148	-13.7626	-0.23555	-0.1727
		6	8.797133	-12.6263		3	112.5	113.3	112.4	113.1		8.825624	-12.715	8.822842	-12.6632	-0.32387	-0.70244
		7	12.48884	-9.42921		4	124.1	103.2	124.1	103.2		12.49506	-9.51249	12.5235	-9.45538	-0.04984	-0.88315
		8	14.82256	-5.51351		4	131.5	90.8	131	90.7		14.83591	-5.58066	14.63627	-5.44422	-0.09001	-1.21785
		9	15.19843	-0.81329		4	132.7	76	137.7	76		15.2155	-0.88783	15.22145	-0.81453	-0.1123	-9.16492
		10	14.00538	4.559365		4	128.9	59.1	131.4	58		14.01345	4.470869	14.74117	4.798896	-0.05758	1.940987
		11	11.46751	9.470605		4	121	43.5	121.7	43		11.51443	9.417362	11.52183	9.515463	-0.40918	0.562198
		12	6.582752	11.45497		1	105.5	37.3	105.5	37.3		6.61131	11.38328	6.581763	11.45325	-0.43384	0.625906
		13	3.227364	12.86578		1	94.9	32.9	94.9	32.8		3.258205	12.77844	3.236061	12.90045	-0.95561	0.678875
		14	-1.13665	13.45259		1	81.1	31	81.1	31		-1.10716	13.3809	-1.1367	13.45326	2.594495	0.53296
		15	-5.52705	12.95829		1	67.3	32.6	67.1	31.9		-5.47252	12.87357	-5.59813	13.12494	0.986689	0.653819
		16	-9.75927	11.6861		1	53.8	36.6	53.8	36.6		-9.74298	11.60523	-9.72848	11.64924	0.16687	0.692
		17	-13.5433	8.719712		2	41.8	45.9	41.7	45.8		-13.539	8.656363	-13.5623	8.732003	0.031785	0.7265
		18	-15.1387	4.296395		2	36.7	59.9	36.1	59.8		-15.1522	4.217202	-15.3519	4.356908	-0.0896	1.843233
		19	-15.7255	-0.25879		2	34.9	74.3	34.7	74.3		-15.7216	-0.34879	-15.774	-0.25959	0.02453	-34.7776
Count	n	19	19	19	n	19	19	19	19	19	n	19	19	19	19	19	19
Sum	Σ	190	0	-1.2E-13	Σ	48	1608.2	1395.2	1614.9	1392.4	Σ	0.253064	-1.39516	0.441465	0.403462	-0.11286	-41.1045
Mean	μ	10	0	-6.3E-15	μ	2.526	84.64211	73.43158	84.99474	73.28421	μ	0.013319	-0.07343	0.023235	0.021235	-0.00594	-2.16339
Population SD	σ	5.477	10.69609	10.32774	σ	1.141	33.83904	32.56152	34.45465	32.67125	σ	10.70432	10.3247	10.7614	10.35959	0.733565	8.009692
MSE																	
ε'_{μ}																	
RMS																	
xLeft			-15.7255														
xRight			15.19843														
yTop				13.45259													
yBottom				-13.8588													

Table 7.7
Boscawen-Ûn pq0 Non-Rotative Method
Correlations of Measured and Computed p and q Points

ELLIPSES (Ground)

	Center		Axes					
Serial	x	y	A	B	a	b	h	k
1	-9.20522	0.380499	13.15935	22.29093	6.579677	11.14546	-9.20522	0.380499
2	0.41123	-10.5589	21.51048	6.753866	10.75524	3.376933	0.41123	-10.5589
3	0.41123	-10.5589	21.51048	6.753866	10.75524	3.376933	0.41123	-10.5589
4	0.41123	-10.5589	21.51048	6.753866	10.75524	3.376933	0.41123	-10.5589
5	0.41123	-10.5589	21.51048	6.753866	10.75524	3.376933	0.41123	-10.5589
6	0.41123	-10.5589	21.51048	6.753866	10.75524	3.376933	0.41123	-10.5589
7	11.70423	-0.09512	7.054173	19.24693	3.527086	9.623466	11.70423	-0.09512
8	11.70423	-0.09512	7.054173	19.24693	3.527086	9.623466	11.70423	-0.09512
9	11.70423	-0.09512	7.054173	19.24693	3.527086	9.623466	11.70423	-0.09512
10	11.70423	-0.09512	7.054173	19.24693	3.527086	9.623466	11.70423	-0.09512
11	11.70423	-0.09512	7.054173	19.24693	3.527086	9.623466	11.70423	-0.09512
12	-1.70819	9.82957	18.53697	7.261198	9.268487	3.630599	-1.70819	9.82957
13	-1.70819	9.82957	18.53697	7.261198	9.268487	3.630599	-1.70819	9.82957
14	-1.70819	9.82957	18.53697	7.261198	9.268487	3.630599	-1.70819	9.82957
15	-1.70819	9.82957	18.53697	7.261198	9.268487	3.630599	-1.70819	9.82957
16	-1.70819	9.82957	18.53697	7.261198	9.268487	3.630599	-1.70819	9.82957
17	-9.20522	0.380499	13.15935	22.29093	6.579677	11.14546	-9.20522	0.380499
18	-9.20522	0.380499	13.15935	22.29093	6.579677	11.14546	-9.20522	0.380499
19	-9.20522	0.380499	13.15935	22.29093	6.579677	11.14546	-9.20522	0.380499
n	19	19	19	19	19	19	19	19
Σ	190	15.2155	-2.60008	288.1456	255.4737	144.0728	127.7368	15.2155	-2.60008
μ	10	0.800816	-0.13685	15.16556	13.44598	7.582778	6.722992	0.800816	-0.13685
σ	5.4772	7.35255	7.401345	5.634456	6.868567	2.817228	3.434284	7.35255	7.401345

Table 7.8
Boscawen-Ûn pq0 Non-Rotative Method
Ellipse Parameters in Ground Meters

Serial	GRADIENTS		POLAR		
	yBos/xBos Gradient	Intersection Gradient	Radius	Circuit Radius Angle	
	g	g'	R	ω	θdeg
1	0.64981914	0.649819139	15.7432127	-2.5653446	-146.98342
2	1.13335098	1.133350979	15.1500642	-2.293768	-131.42322
3	2.37008666	2.370086665	14.4908591	-1.97006	-112.87612
4	20.1129409	20.11294086	13.8759556	-1.6204747	-92.846358
5	-3.6108253	-3.61082534	14.1890803	-1.3006228	-74.520196
6	-1.435278	-1.43527799	15.3887556	-0.9622689	-55.133949
7	-0.7550114	-0.7550114	15.6486758	-0.6467007	-37.05322
8	-0.3719675	-0.37196752	15.8147794	-0.3561094	-20.403566
9	-0.0535117	-0.05351172	15.2201797	-0.0534607	-3.063074
10	0.32554382	0.325543818	14.7288332	0.31472368	18.0323389
11	0.82586404	0.825864039	14.8726656	0.69031394	39.5520756
12	1.74014969	1.740149691	13.2117003	1.04921519	60.1156024
13	3.98646755	3.986467546	13.264398	1.32501909	75.9180019
14	-11.835328	-11.8353279	13.5005287	1.65508892	94.8296098
15	-2.3445199	-2.34451987	14.0877814	1.97395928	113.099535
16	-1.1974362	-1.19743624	15.2252515	2.26658665	129.865849
17	-0.6438416	-0.64384164	16.1075488	2.56955888	147.224879
18	-0.2838026	-0.28380262	15.7365313	2.86506127	164.155919
19	0.01645673	0.016456728	15.7276183	-3.1251374	-179.05718
n	19	19	19	19	19
Σ 190	8.62915727	8.629157266	281.984419	-0.1844202	-10.5665
μ 10	0.45416617	0.454166172	14.8412852	-0.0097063	-0.5561316
σ 5.4772	5.57699108	5.576991081	0.89709344	1.79613882	102.911174

Table 7.9
Boscawen-Ûn pq0 Non-Rotative Method
Gradients and Polar Ray Statistics

AUXILIARIES

Serial	φ	μ	$b^2+a^2g^2$	sgnX	sgnY	U	V	W	X	Y
1	-0.3804995	-5.9817287	142.502111	-1	-1	-1132.7808	-740.72025	142.502111	-736.10264	-481.33419
2	10.5588602	0.4660677	159.986678	-1	-1	-1379.5835	-225.1754	159.986678	-1563.5523	-255.20276
3	10.5588602	0.97465026	661.187264	-1	-1	-2890.1304	-834.69446	661.187264	-6849.8595	-1978.2982
4	10.5588602	8.27104063	46805.5402	-1	-1	-24561.223	-7827.8045	46805.5402	-493998.42	-157440.17
5	10.5588602	-1.484879	1519.58392	1	1	4414.94558	1376.91848	1519.58392	-15941.597	-4971.8121
6	10.5588602	-0.5902291	249.69725	1	1	1757.73567	445.303597	249.69725	-2522.8393	-639.13445
7	0.09512487	-8.8368291	99.7026088	1	1	1084.83523	163.790129	99.7026088	-819.06297	-123.66341
8	0.09512487	-4.3535944	94.3323341	1	1	1084.38194	296.291182	94.3323341	-403.35486	-110.2107
9	0.09512487	-0.6263136	92.6467139	1	1	1084.00509	326.212244	92.6467139	-58.006972	-17.456177
10	0.09512487	3.81024061	93.9295028	1	1	1083.55652	301.07397	93.9295028	352.745126	98.0127695
11	0.09512487	9.66610494	101.096041	1	1	1082.96445	81.846726	101.096041	894.381394	67.5942677
12	-9.8295695	-2.9724982	273.31153	1	1	1446.8798	351.991925	273.31153	2517.78744	612.51864
13	-9.8295695	-6.8096253	1378.37458	1	1	3343.68792	1116.81621	1378.37458	13329.5034	4452.15158
14	-9.8295695	20.2169333	12046.3021	1	1	-10016.358	-3676.7177	12046.3021	118546.882	43515.1596
15	-9.8295695	4.00487442	485.380753	-1	-1	-2002.2467	-714.97974	485.380753	4694.30709	1676.2842
16	-9.8295695	2.04544302	136.356227	-1	-1	-1033.6404	-292.8987	136.356227	1237.7185	350.727518
17	-0.3804995	5.92670449	142.167339	-1	-1	-1154.0908	-774.03186	142.167339	743.051709	498.353937
18	-0.3804995	2.61246583	127.708276	-1	-1	-1148.16	-812.40399	127.708276	325.850821	230.562384
19	-0.3804995	-0.1514878	124.233081	-1	-1	-1143.2139	-816.4434	124.233081	-18.81356	-13.435987
n	19	19	19	19	19	19	19	19	19	19
Σ	190	2.60007968	26.18734	64734.0386	-1	-1	-30078.435	-12255.626	64734.0386	-380269.39	-114529.35
μ	10	0.1368463	1.37828105	3407.05466	-0.0526316	-0.0526316	-1583.0755	-645.03293	3407.05466	-20014.178	-6027.8607
σ	5.4772	7.40134545	6.50235965	10564.3426	0.998614	0.998614	6148.46287	1980.0322	10564.3426	114936.467	37023.0514

Table 7.10
Boscawen-Ûn pq0 Non-Rotative Method
Auxiliaries

Radii						
Serial	R_{bos}	r_q	$R_{bos}-r_q$	R_{bos}/r_q	$(R_{bos}-r_q)^2$	$(R_i)^2$
1	15.7432127	15.6791635	0.06404921	1.00408499	0.0041023	247.848745
2	15.1500642	15.1607283	-0.0106641	0.9992966	0.00011372	229.524445
3	14.4908591	14.4917936	-0.0009345	0.99993552	8.7327E-07	209.984998
4	13.8759556	13.9351725	-0.0592169	0.99575055	0.00350664	192.542144
5	14.1890803	14.2806249	-0.0915446	0.99358959	0.00838042	201.329999
6	15.3887556	15.4337271	-0.0449716	0.99708615	0.00202244	236.813797
7	15.6486758	15.6921082	-0.0434324	0.99723222	0.00188637	244.881056
8	15.8147794	15.6160098	0.19876953	1.01272857	0.03950932	250.107246
9	15.2201797	15.2432279	-0.0230482	0.99848797	0.00053122	231.65387
10	14.7288332	15.5026258	-0.7737926	0.95008635	0.59875498	216.938527
11	14.8726656	14.9431106	-0.070445	0.99528579	0.0049625	221.196182
12	13.2117003	13.2097164	0.00198388	1.00015018	3.9358E-06	174.549024
13	13.264398	13.300141	-0.035743	0.99731258	0.00127756	175.944253
14	13.5005287	13.5011969	-0.0006682	0.99995051	4.4651E-07	182.264274
15	14.0877814	14.2689539	-0.1811726	0.98730302	0.0328235	198.465584
16	15.2252515	15.1772227	0.0480288	1.00316453	0.00230677	231.808283
17	16.1075488	16.1302545	-0.0227057	0.99859235	0.00051555	259.453127
18	15.7365313	15.9581739	-0.2216427	0.98611102	0.04912548	247.638416
19	15.7276183	15.7761739	-0.0485556	0.99692222	0.00235764	247.357978
n	19	19	19	19	19	19	19
Σ	190	281.984419	283.300125	-1.3157061	18.9130707	0.75218166	4200.30195
μ	10	14.8412852	14.9105329	-0.0692477	0.99542478	0.03958851	221.068524
σ	5.477	0.89709344	0.89769228	0.18652953	0.01204268	0.13257097	26.2627941

MSE	0.03958851
ε'_μ	0.00017908
RMS	0.19896861

Table 7.11
Boscawen-Ûn pq0 Non-Rotative Method
Radii And Error

Polar Profile Group Serial	θ	R_{bos}	r_q
1	-2.56534	15.74321	15.67916
2	-2.29377	15.15006	15.16073
3	-1.97006	14.49086	14.49179
4	-1.62047	13.87596	13.93517
5	-1.30062	14.18908	14.28062
6	-0.96227	15.38876	15.43373
7	-0.6467	15.64868	15.69211
8	-0.35611	15.81478	15.61601
9	-0.05346	15.22018	15.24323
10	0.314724	14.72883	15.50263
11	0.690314	14.87267	14.94311
12	1.049215	13.2117	13.20972
13	1.325019	13.2644	13.30014
14	1.655089	13.50053	13.5012
15	1.973959	14.08778	14.26895
16	2.266587	15.22525	15.17722
17	2.569559	16.10755	16.13025
18	2.865061	15.73653	15.95817
19	-3.12514	15.72762	15.77617

n	19	19	19	19
Σ	190	-0.18442	281.9844	283.3001
μ	10	-0.00971	14.84129	14.91053
σ	5.477226	1.796139	0.897093	0.897692

Table 7.12
Boscawen-Ûn pq0 Non-Rotative Method
Polar Profile Group

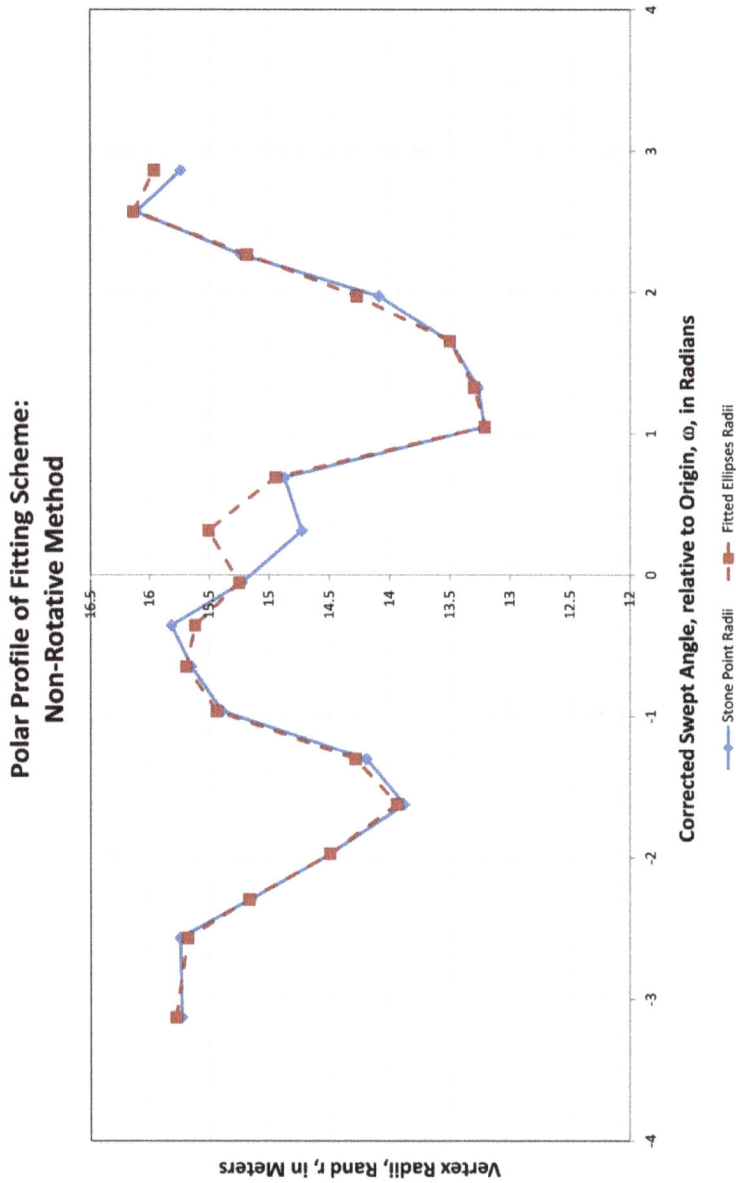

Figure 7.2
The Boscawen-Ûn Polar Profile of R_i with r_i
For the Non-Rotative Ellipse Fitment Method

CHAPTER EIGHT
COMPARISONS AND INTERPRETATIONS

I had initially intended to call this Chapter "Conclusions" but in science of course there are never conclusions: Only surprise discoveries and provisional, nay, tentative gleanings, often as beautiful and as innocent as Newton's seashell on his beach.

The statistics are most suggestive that we were right to think that the Ancients drafted the Boscawen-Ûn stone circuit upon their land by dragging taut ropes round turnposts to describe four ellipses.

But at the same time the statistics define a dramatic difference in the quality of the elliptical fitments between our Rotated and Non-Rotated correlations. The Non-Rotated process is dramatically better.

Table 8.1 shows the 22.37-fold superiority of the Non-Rotated Method:-

Name	Symbol	Rotative Fitment Method	Non-Rotative Fitment Method
Mean Square Error	MSE	0.88569655	0.03958851
Relative Error Fraction	ε'_μ	0.00400643	0.00017908
Root Mean Square Error	RMS	0.94111452	0.19896861

Table 8.1
A Comparison of the Rotative and Non-Rotative Treatments in
Ellipses Fitment to the Boscawen-Ûn Stone Circuit Data

We expect technical complications to increase numerical error, but the scope of the error surprised at least this researcher.

We have seen how useful Polar Profile graphs are in the qualitative comparison of data fitments, and I invite you to reprise them with the following two diagrams, Figure 8.1 and 8.2, respectively for the Rotative and Non-Rotative fitment methods.

You can easily perceive how neatly the red dashed Fitted Ellipse Radii fit the blue continuous Stone Position Radii in the Non-Rotative case

.

The Eight Ellipse Foci

Each of the four fitted ellipses obviously have two foci (none are circles). We need to establish the positions of each of the eight foci and if possible make an assessment of the putative square or rectangle at whose corners the Ancients may have driven their turnposts.

As a passing check we can compare the Local against the Global Graphical Calibration Constants as shown in Table 8.2

From there we can move forward to establish the all-important Focal Distances from the relevant Sector Ellipses' Centers using the equation:-

$$\pm f = \sqrt{a^2 - b^2}$$
Equation 8.1

where f is the Axial Distance from The Focus to the Ellipse Center; a is the Ellipse Major Half-Axis Length; and b is the Ellipse Minor Half-Axis Length.

These f-displacements are quoted in Table 8.3

Once we have these measurements we may supplement the Veusz scale plot of the Boscawen-Ûn stone circuit to show the foci in case they throw light upon the layout practices of the Ancient builders. The eight foci are shown as empty hard black circles in Figure 8.3

Except in the NE corner where the Eastern focus of Sector One is separated from the Northern of Sector Four by some three meters, the other corner foci are within roughly 1.2 meters of one another. This precludes the possibility of other than a tetrapole conformation, and in particular it precludes an octopole. It virtually certifies a rectangular closed-loop geoscriptive system that is a practical, as opposed to a theoretical, tetrapole.

The existing focus separation at the NW, SW and SE corners is very consistent in both azimuth and length and could simply reflect the stoutness of the turnposts sunk, perhaps the thick trunks of mature broad-leaved trees. Or alternatively these discrepancies as well as the much wider NE gap may wholly or partially reflect land-creep occurring in the intervening millennia.

Land-creep is a geological phenomenon involving the mass-slippage of topsoil and unindurated surficial sediments. This is occasioned by all or some of soil solifluction, freeze-thaw, glacial disturbances, landslip, and interference by plants and animals, including human activity.

Note the fact that the extrapolated axes of the fitted ellipses tend to describe a rough square on the ground between the four corner focus clusters.

Table 8.4 presents the dimensions of this ground figure.

I am satisfied that the Ancients drew the Boscawen-Ûn circuit around a square of some twenty meters side using a closed loop of hide or rope. any departure from a precise square is due to the vagaries of time. I hope and expect that non-destructive surveys, for example Ground-Penetrating Radar, shall locate the four ancient postholes.

The sum of the four sides of the square represents the lower limit of the Loop Length, but the length of a useful loop would not be markedly longer. Table 8.4 Presents the Sum of Edges as 82.65(9108) meters. This is consistent with the previously-determined L-values of 82.42296006 for the North and South ellipses, and 76.08273236 for the West and East.

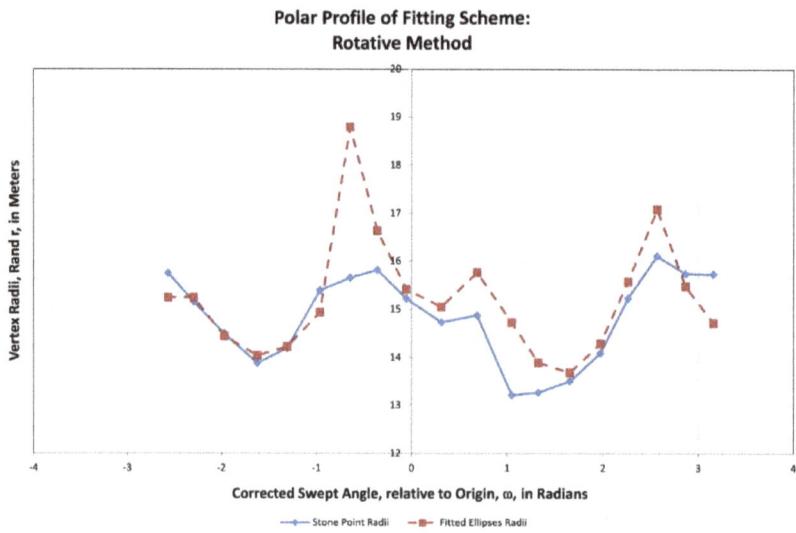

Figure 8.2
Polar Profile for the Rotative Ellipses Fitment Method

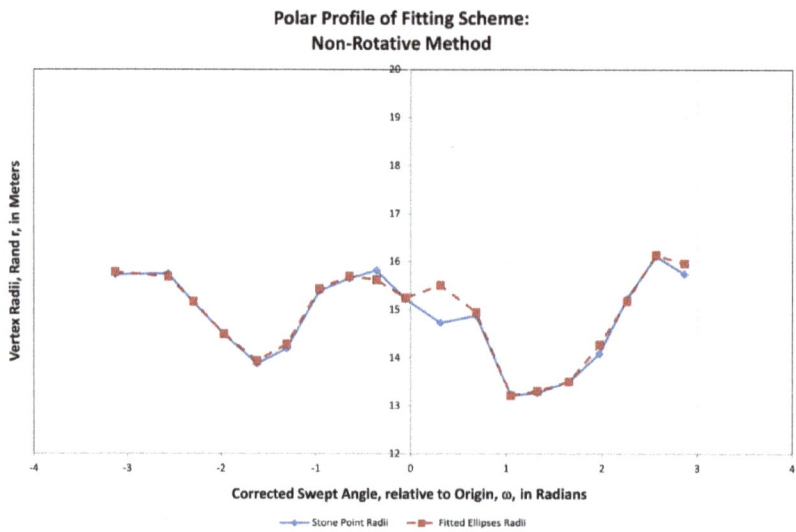

Figure 8.2
Polar Profile for the Rotative Ellipses Fitment Method

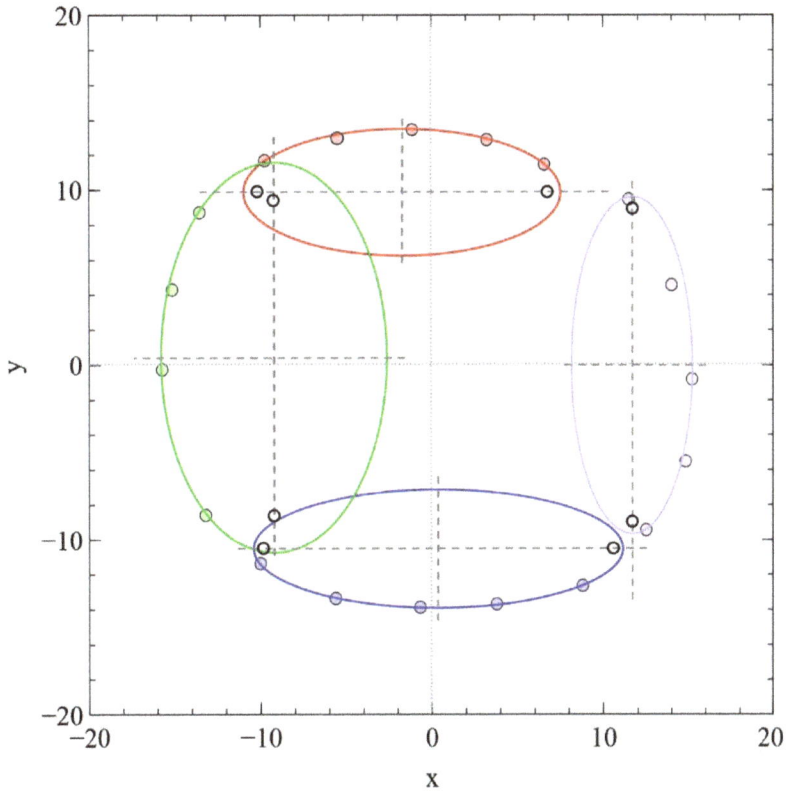

Figure 8.3
The Scale Plan of the Boscawen-Ûn Stone Circuit
Showing the Eight Ellipse Foci

Corners	Veuze/Ground Plot Distance	PhotoDraw Plot Distance	Local x Calibration Dividend	Local y Calibration Dividend	x Calibration Mean	x Calibration Mean
NW-NE	40	126.5	0.31620553		0.31633061	
NW-SW	40	126.1		0.31720856		0.31708289

Table 8.2
Local and Global Graphical Calibration Constants

| Sector | Ellipse Center | | Implied Focis | | Focal Axis | | Veusz/Photodraw Distance Relative to Ellipse Center | | | |
| | | | | | | | Focus Position, -f | | Focus Position, +f | |
	x	y	-f	+f	x	y	x	y	x	y
1	-1.70818532	9.82956952	-8.52781344	8.52781344	YES	NO	-26.95854601		26.95854601	
2	-9.20522089	0.38049947	-8.99606636	8.99606636	NO	YES		-28.37133984		28.37133984
3	0.41122980	-10.55886016	-10.21134331	10.21134331	YES	NO	-32.28060398		32.28060398	
4	11.70423275	-0.09512487	-8.95381219	8.95381219	NO	YES		-28.23808077		28.23808077

Table 8.3
Focal Distances relative to Ellipse Centers

	PhotoDraw (mm)		Veusz/Ground (meters)	
	x	y	x	y
Corner				
NW	55.5	42.3		
SW	55.5	106.7		
SE	121.7	106.6		
NE	121.6	42.3		
Mean x Edge	66.15		20.93	
Mean y Edge		64.35		20.40
Mean Edge	20.66			
Sum of Edges	82.66			

Table 8.4
The Formative Rectangle

The Problem of Geoscription between the Side-Sectors

I investigated the behavior of a closed loop consistent with the Sector One and Sector Two draftings when it was or might have been dragged between them.

I thought of this as the "Sector 1.5 Problem", or, to satisfy EXCEL® sheet naming conventions, the "Sector 1dot5" problem.

Table 8.5 lists the generative parameters of the elliptical arcs which fit Sectors One, Two, and the intermediate Sector 1.5 for this second or third exercise of hand-and-eye fitments. The actual graphical arcs were elaborated as sixty-four line segments interpolated between the angles ω_0 and ω_n for $n_{inc} = 64$. Note that in the Plot of Figures which is Figure 8.4 the range of ω is $0 \leq \omega \leq \pi$ whilst it was found that $-10/13 \leq \omega \leq +10/13$ was adequate to illustrate both the Sector One and Sector Two arcs of approximation.

Table 8.6 presents the formulae underlying these outcomes in the idiom of EXCEL®.

The A and B values for the two fitted ellipses of respectively Sectors One and Two were established heuristically. They are very similar in their values, suggesting that they were originally identical before "land-creep" distorted the circuit figure.

The aspect of the intermediate Sector 1.5 ellipse (if any) was of course unknown, but I decided to assume that it was a simple mean of the respective Sector One and Two Major and Minor Axes.

In each Sector the Foci of the Fitted Ellipses is given by Equation 8.1 but for Sector 1.5 I assumed the relevant ellipse foci to reside at $D_{S1.5}/2$ (i.e. the SW and NE turnposts) where $D_{S1.5}$ is given by:-

$$D_{S1.5} = \sqrt{2f_{S1}{}^2 + 2f_S{}^2} = \sqrt{2\left(f_{S1}{}^2 + f_S{}^2\right)}$$

Equation 8.2

Recall that a key postulate is that parts of our closed loop turn about ellipse foci. Therefore, we may identify sides of the tetrapole with ellipse interfocal distances 2f in this manner:-

$$LC_{S1} = 2f_{S1} + 4f_{S2}$$
Equation 8.3

$$LC_{S2} = 2f_{S2} + 4f_{S1}$$
Equation 8.4

$$LC_{S1.5} = 2f_{S1} + 2f_{S2}$$
Equation 8.5

where for the given Sector LC_S is the Length of the Loop in Contact with Sides of the Turnpost Rectangle.

Name	Symbol	SECTOR ONE Value	SECTOR TWO Value	SECTOR 1.5 Value
MAJOR AXIS:	A	28.000000	27.000000	27.500000
MINOR AXIS:	B	16.500000	16.386859	16.443429
MAJOR SEMI-AXIS:	a	14	13.5	25.91604038
MINOR SEMI-AXIS:	b	8.25	8.193429427	20.70231713
NEGATIVE FOCUS:	-f	-11.31094603	-10.72929234	-15.59022816
POSITIVE FOCUS:	+f	11.31094603	10.72929234	15.59022816
DIAGONAL:	D	32.5	31.58368476	31.18045633
ECCENTRICITY:	e	0.807924716	0.794762396	0.601566749
TILT (radians):	α	0	4.71238898	-0.811782574
TILT (degrees):	α^o	0	270	-46.51171539
X-SHIFT:	h	-2	-7.25	0
Y-SHIFT:	k	5	0	0
FULL ELLIPSE AREA:	K	362.8539515	347.4956389	1685.533846
NOMINAL MEAN RADIUS:	r	10.74709263	10.51719056	23.16294641
ARC FRACTION:	γ	1	1	0.769230769
ARC HALF-ANGLE:	θ	0.785398163	0.785398163	0.655695626
SWEPT ANGLE ω LOWER BOUND:	ω_0	0	0	0.915100701
SWEPT ANGLE ω UPPER BOUND:	ω_n	3.141592654	3.141592654	2.226491953
SWEPT ANGLE ω INCREMENT:	ω_{inc}	0.049087385	0.049087385	0.020490488
NUMBER OF INCREMENTS:	n_{inc}	64	64	64
FREE LOOP LENGTH:	LF_s	28	31.58368476	51.83208075
CONTACT LOOP LENGTHI:	LC_s	65.53906143	66.70236879	44.08047674
MINIMUM LOOP LENGTH:	LL_s	93.53906143	98.28605355	95.91255749

Table 8.5
Sector Parameter Values

Name	Symbol	SECTOR ONE Value	SECTOR TWO Value	SECTOR 1.5 Value
MAJOR AXIS:	A	28	27	=('SECTOR ONE'!$C$2+'SECTOR TWO'!$C$2)/2
MINOR AXIS:	B	16.5	16.386858854714	=('SECTOR ONE'!$C$3+'SECTOR TWO'!$C$3)/2
MAJOR SEMI-AXIS:	a	=C3/2	=D3/2	=SQRT(E8^2+E6^2)
MINOR SEMI-AXIS:	b	=C4/2	=D4/2	=SQRT((E26/2)^2-E8^2)
NEGATIVE FOCUS:	-f	=-C8	=-D8	=-E8
POSITIVE FOCUS:	+f	=SQRT(C5^2-C6^2)	=SQRT(D5^2-D6^2)	=E9/2
DIAGONAL:	D	=SQRT(C3^2+C4^2)	=SQRT(D3^2+D4^2)	=SQRT((2*'SECTOR ONE'!$C$7)^2+(2*'SECTOR TWO'!$C$7)^2)
ECCENTRICITY:	e	=SQRT(1-C6^2/C5^2)	=SQRT(1-D6^2/D5^2)	=SQRT(1-E6^2/E5^2)
TILT (radians):	α	0	=3*PI()/2	=-PI()/2+ATAN2(2*'SECTOR ONE'!$C$7,2*'SECTOR TWO'!$C$7)
TILT (degrees):	$\alpha°$	=180*C11/PI()	=180*D11/PI()	=180*E11/PI()
X-SHIFT:	h	-2	-7.25	0
Y-SHIFT:	k	5	0	0
FULL ELLIPSE AREA:	K	=PI()*C5*C6	=PI()*D5*D6	=PI()*E5*E6
NOMINAL MEAN RADIUS:	r	=SQRT(C16/PI())	=SQRT(D16/PI())	=SQRT(E16/PI())
ARC FRACTION:	γ	1	1	10/13
ARC HALF-ANGLE:	θ	=ATAN(C19)	=ATAN(D19)	=ATAN(E19)
SWEPT ANGLE ω LOWER BOUND:	ω_0	0	0	=-E20+PI()/2
SWEPT ANGLE ω UPPER BOUND:	ω_n	=PI()	=PI()	=E20+PI()/2
SWEPT ANGLE ω INCREMENT:	ω_{inc}	=(C22-C21)/C24	=(D22-D21)/D24	=(E22-E21)/E24
NUMBER OF INCREMENTS:	n_{inc}	64	64	64
FREE LOOP LENGTH:	LF_S	=2*(SQRT(C6^2+C7^2))	=2*(SQRT(D5^2+D6^2))	=E28-E27
CONTACT LOOP LENGTH:	LC_S	=2*C8+4*'SECTOR TWO'!$C$7	=2*D8+4*'SECTOR ONE'!$C$7	=2*'SECTOR ONE'!$C$7+2*'SECTOR TWO'!$C$7
MINIMUM LOOP LENGTH:	LL_S	=C26-C27	=D26-D27	=('SECTOR ONE'!$C$27+'SECTOR TWO'!$C$27)/2

Table 8.6
Sector Parameter Formulae

Also, LF$_S$ is the Sector Free Loop Length (i.e., the part of the loop *not* extending about the turnposts) and is given by:-

$$LF_{S1} = 2\sqrt{f_{S1}^2 + b^2}$$
Equation 8.6

$$LF_{S2} = 2\sqrt{f_{S2}^2 + b^2}$$
Equation 8.7

$$LF_{S1.5} = \frac{LL_{S1} + LL_{S2}}{2} - 2(f_{S1} + f_{S2})$$

Equation 8.8

From which it follows that the Minimum Loop Length, LLS, is given by:-:-

$$LL_S = LC_S + LF_S$$
Equation 8.9

Or for the case of Sector 1.5 we may as well average the Sector Oner and Sector Two LL values as:-

$$LL_{S1.5} = \frac{LL_{S1} + LL_{S2}}{2}$$
Equation 8.10

It is most probable that the Ancient Britons who laid-out Boscawen-Ûn stone circuit thought that the sensible choice of loop length was 95±5 meters. At any event, we have no reason to confute them.

Figure 8.4 is an EXCEL® plot representing the Figure of Plan of Boscawen-Ûn. Each of the nineteen circuit monoliths is represented by a light gray triangle and a dashed gray line is included to guide the eye: There is no implication that the gray trace was emplaced by Ancient planners or architects. The heavy red line is a 180-degree trace of the partial ellipse fitted by eye to the orthostats of Sector One. Similarly, the green dot-dash trace is fitted to the stones of Sector Two. The black rectangle represents the nominal figure of the tetrapole, whose corner turnposts are represented schematically by large black circles. The black circles are foci but no attempt has been made to make this black structure conform to the shift in the position of the stones. The dashed black diagonal is the Interfocal Base $2f_{S1.5}$ of the Intermediate Ellipse S1.5 whose geoscriptive arc is shown by the continuous orange line standing some distance North-West of the megalithic circuit.

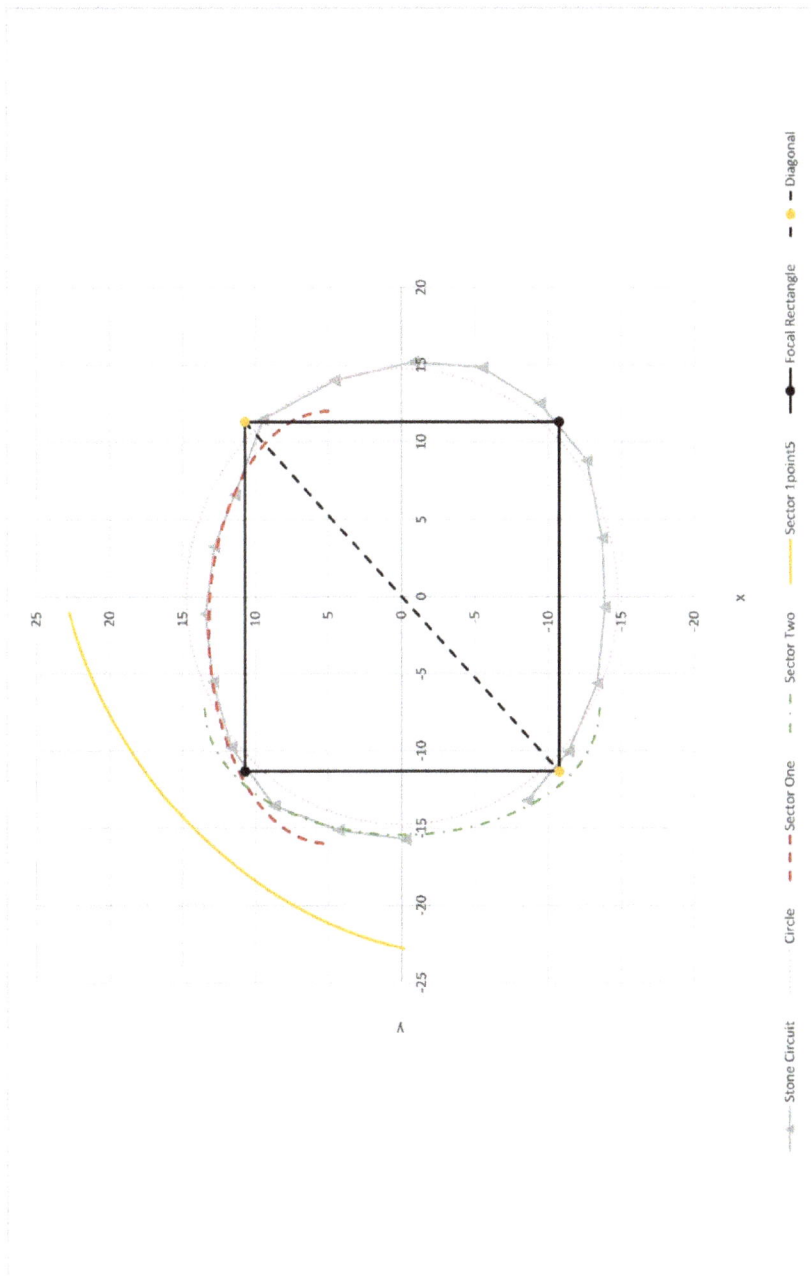

Figure 8.4
Boscawen-Ûn Plot of Figures

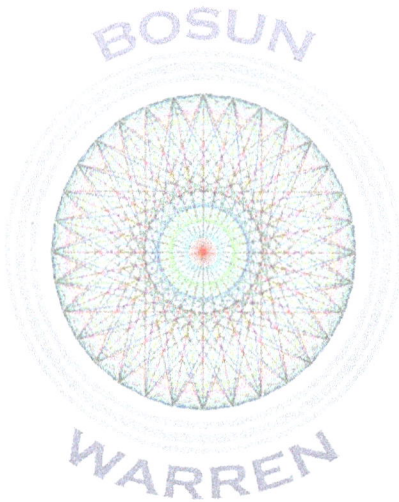

CHAPTER NINE
REFERENCES

CHAPTER ZERO

(no references)

CHAPTER ONE

1.1 "Some Design Aspects of Roman Encampments in Britain:
An Extended Study"
James R Warren
18 February 2013

1.2 **Ancient Egyptian Calibrated Ropes and Megalithic Metrology**
Wikipedia contributors. (2021, December 4).
Rope stretcher.
In Wikipedia, The Free Encyclopedia.
Retrieved 12:03, December 26, 2021, from
https://en.wikipedia.org/w/index.php?
title=Rope_stretcher&oldid=1058565819

"Megalithic Metrology and Design
An Alternative Route to the Megalithic Yard"
GJ Bath
http://www.gjbath.com/Stones/MegalithicUnit.htm

"The Design and Distribution of Stone Circles
in Britain; a Reflection of Variation in
Social Organization in the Second and Third
Millennia BC."
A thesis submitted for the Degree of Doctor of
Philosophy in the Department of Archaeology
and Prehistory, University of Sheffield.
December 1987.
by

John Barnatt.
pp329
https://etheses.whiterose.ac.uk/14854/

1.3   **Ancient Egyptian Calibrated Ropes**
EGYPTIAN SURVEYING TOOLS
by Mary M. Root
Article taken from "Backsights" Magazine published
by Surveyors Historical Society
http://www.surveyhistory.org/
egyptian_surveying_tools1.htm

1.4   **Wikimedia Commons "Boscawen-Ûn"**
File:01 Boscawen-un.JPG. (2020, September 6).
Wikimedia Commons, the free media repository.
Retrieved 11:53, February 25, 2022 from
https://commons.wikimedia.org/w/index.php?
title=File:01_Boscawen-un.JPG&oldid=449523472.

1.5   01_Boscawen-un.jpg
4032×3024 3.6MB
Revision as of 03:25, 7 January 2019
 by YiFeiBot (talk | contribs)
(Bot: Migrating 2 langlinks, now
 provided by Wikidata on d:q894360)
(diff) ← Older revision | Latest revision (diff) |
Newer revision → (diff)
By Waterborough - Own work, CC BY-SA 3.0,
https://commons.wikimedia.org/w/index.php?
curid=75349112
https://commons.wikimedia.org/wiki/
Category:Boscawen-Un

CHAPTER TWO

2.1   **Side-by-Side 25 inch to one mile GBOS Maps (1892-1914)**

The National Libraries of Scotland
Side-by-Side 25 inch to one mile GBOS Maps (1892-1914)
Collection with Aerial Photograph
ESRI World Imagery

https://maps.nls.uk/geo/explore/side-by-side/#zoom=19&lat=50.08992&lon=-5.61931&layers=168&right=ESRIWorld

2.2   **Ragazzo Oval**
"All Sides to an Oval"
"Properties, Parameters, and Borromini's Mysterious Construction"
Angelo Alessandro Mazzotti
ISBN 978-3-319-81879-5
DOI 10.1007/978-3-319-39375-9
© Springer International Publishing AG of Cham
pp160

2.3   **Cubic Superellipse**
Wikipedia contributors. (2021, November 23).
Superellipse.
In Wikipedia, The Free Encyclopedia.
Retrieved 14:27, December 26, 2021, from
https://en.wikipedia.org/w/index.php?title=Superellipse&oldid=1056710377

CHAPTER THREE

3.1   **Perimeter of the General Ellipse**
Wikipedia contributors. (2021, December 21).
Ellipse.
In Wikipedia, The Free Encyclopedia.

Retrieved 14:33, December 26, 2021, from
https://en.wikipedia.org/w/index.php?
title=Ellipse&oldid=1061346573

CHAPTER FOUR
(no references)

CHAPTER FIVE

5.1   Ellipse Line Intersection Calculator
AmBrSoft Calculators
Ellipse line intersection (ambrsoft.com)
http://www.ambrsoft.com/TrigoCalc/Circles2/
Ellipse/EllipseLine.htm

5.2   "Handbook of Mathematical Formulas"
Hans-Jochen Bartsch
Translated from the German by Herbert Liebscher
Ninth Edition: 1974
© VEB Fachbuchverlag Leipzig 1974
Academic Press of New York and London
        A Subsidiary of Harcourt Brace Jovanovich
        Publishers
Library of Congress Catalog Card Number:
73-2088
ISBN 0-12-080050-0
pp525 Hardback
Rigid Rotation in a Plane   p215
Intersections of an Ellipse   p235

THE NATIONAL LIBRARY OF SCOTLAND

Map figures, including NLS Side-by-Side images, are:-

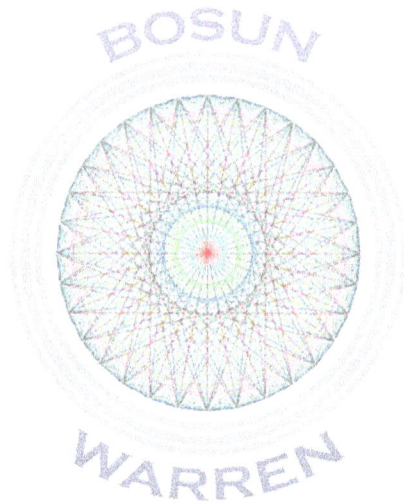

INDEX

carbon-dating, 10

Carn Euny, 24

Cartesian, 29, 32, 34, 53, 55, 70, 80, 103

Catalog, 140

celestial, 18

Cell, 80, 82

Celtic, 24

Celts, 10

center, 24, 30, 38, 39, 43, 52, 63, 78, 86, 89, 99

Center, 36, 39, 45, 61, 63, 70, 89, 110, 122

chain, 22

Cheshire, 11

Christians, 11

circle, 11, 13, 22, 23, 25, 38, 43, 49, 57, 58, 61, 62, 63

Circle, 19, 20, 46, 49, 53, 56, 58, 61, 62, 67, 68

circuit, 18, 21, 25, 28, 29, 31, 36, 37, 38, 40, 44, 45, 49, 57, 58, 61, 69, 71, 74, 86, 87, 99, 102, 121, 122, 123, 129, 133

circumference, 11

Circumference, 13, 36, 50

classical, 48

Classical, 11

clocks, 11

clouds, 86

Code, 2

Coefficient, 31

coins, 11

Combinations, 62

commons, 138

compactness, 38

Comparisons, 49

Complete, 47

complex, 16, 22, 72, 86

Conservation of Area, 44

Constant, 11, 13, 37, 89

construction, 11, 21, 49, 86, 87

Contents, 7

coordinates, 30

Coordinates, 110

Copper, 10

Copyright, 2

Cornish, 16, 19, 21, 23, 24, 104

Cornwall, 16, 24

Corrected, 37, 39, 46

cos, 70, 71, 72, 73

creeks, 16

Criticism, 86

Cross, 2, 24

Crows-an-Wra, 24

crushing, 43

Cubic, 28, 139

curves, 22

Cyclopean, 9

Dark Age, 24

Defect, 14, 40, 51

degree, 11, 25, 44, 133

Degrees of Freedom, 86

Delaunay, 30

dendrochronology, 10

Design, 2, 22, 45, 137

deviation, 62, 86

Deviation, 14, 40, 79

diameter, 11

Diameter, 13

Digitisation, 29

dimensionless, 38, 41, 45

Ding-Dong, 24

dipole, 58

displacement, 15

distribution, 38
drawing, 14, 50, 71
eccentrically-placed, 18
Eco, 43
Edition, 2, 3, 140
Egyptian, 21, 137, 138
electrons, 43
ellipse, 22, 23, 47, 49, 51, 52, 58, 69, 70, 71, 72, 74, 75, 78, 86, 96, 99, 100, 101, 102, 103, 112, 129, 133
Ellipses, 69, 74, 96, 107, 108, 121, 122, 124, 129
ellipsoid, 43
Elliptic, 47
Encampments, 22, 137
envelope, 44, 45, 46, 51, 58
environments, 10
Error, 40, 45, 49, 58, 75, 79, 91, 118
Euclidean, 30, 38
Europe, 10, 11
EXCEL, 32, 34, 46, 53, 79, 80, 81, 82, 129, 133
experimental, 21
Extended, 22, 137
fabric, 43
Farm, 16
Figure, 19, 20, 23, 27, 28, 29, 31, 32, 34, 36, 38, 40, 46, 53, 54, 56, 57, 58, 63, 67, 68, 69, 71, 73, 74, 76, 77, 78, 80, 86, 99, 100, 101, 113, 120, 121, 122, 124, 125, 129, 133, 135
File, 2, 138
Fitment, 74, 76, 96, 120, 121, 124
Fitments, 101

fitted, 44, 45, 51, 62, 99, 100, 103, 104, 122, 123, 129, 133
foci, 23, 52, 86, 122, 129, 134
Focus, 70, 122
Formulas, 140
fraction, 11, 15, 21, 79
France, 10
freeze-thaw, 87, 123
French, 16, 24
general, 86
GeoHack, 18
Geometric, 13, 110
Geometry, 76
geophysical, 21
geoscribe, 77
geoscription, 86, 87
Giza, 21
gnomon, 18, 19, 23, 27, 28, 29
gnomon-like, 18
God, 5, 9
gorse, 24
Gradients, 110, 116
granite, 10, 16, 18, 23, 24
Greater, 13
Greece, 11
Greeks, 10, 14, 22
Greenwich, 27
Grid, 18, 27
Ground, 103, 109, 115, 123
Half-Axis, 122
Handbook, 140
Harcourt, 140
Headland, 24
hemp, 22
henge, 14
Heron, 37

www.ingramcontent.com/pod-product-compliance
Lightning Source LLC
Chambersburg PA
CBHW041602260326

41914CB00011B/1353